JUMP TIME

ALSO BY JEAN HOUSTON

A Passion for the Possible

A Mythic Life: Learning to Live Our Greater Story

Life Force: The Psycho-Historical Recovery of the Self

The Possible Human

The Search for the Beloved: Journeys in Sacred Psychology

Erwachen (in German)

Godseed: The Journey of Christ

The Hero and the Goddess: The Odyssey as Mystery and Initiation

Public Like a Frog: Entering the Lives of Three Great Americans

The Passion of Isis and Osiris

Manual for the Peacemaker: An Iroquois Legend to Heal Self & Society (with Margaret Rubin)

COAUTHORED WITH ROBERT MASTERS

The Varieties of Psychedelic Experience

Psychedelic Art

Mind Games

Listening to the Body

COAUTHORED WITH DIANA VANDENBERG

A Feminine Myth of Creation

JUMP TIME

SHAPING YOUR FUTURE IN A WORLD OF RADICAL CHANGE

JEAN HOUSTON

SENTIENT PUBLICATIONS, LLC

First Sentient Publications edition, 2004

Book design by Trish Broersma

Library of Congress Cataloging-in-Publication Data

Houston, Jean.
Jump time : shaping your future in a world of radical change / Jean Houston.-- 2nd ed.
p. cm.
ISBN 1-59181-018-3
1. Consciousness. 2. Social evolution. I. Title.
BF311.H653 2004
303.49'09'05--dc22

2004005794

Printed in the United States of America

10 9 8 7 6 5 4 3 2 1

SENTIENT PUBLICATIONS
A Limited Liability Company
1113 Spruce St.
Boulder, CO 80302
www.sentientpublications.com

ACKNOWLEDGMENTS

For many years I have talked about this most amazing time that demands that we "Jump!" to a new dispensation of humanity. One deep and conscientious friend not only listened to me say again and again "I need to write about this!" but also took me seriously and believed that what I had to say could be wrested into a context both useful and interesting. Brenda Rosen is that friend. Brenda is a wondrously innovative editor and a pattern-seeker extraordinaire. Her capacity to focus, persist, evoke, elucidate, and question kept me exploring the array of issues that *Jump Time* presents. I could not have embraced these complexities without her clarity of vision. I certainly could not have written this book without her belief in its possibilities and the hard work she invested in its editing and completion.

I am also indebited to Joel Fotinos, the publisher of Tarcher/Putnam, who was willing to take a chance on this book when it was still in an embryonic state. His wise counsel helped shape and clarify the book's thrust and form. It was a pleasure to work with a man so gifted in mind and spirit. I am also grateful to Mitchell Horowitz, senior editor at Tarcher/Putnam. His advice kept me from falling into various sinkholes, stews, and silliness. And, of course, my old friend Jeremy Tarcher's consistent belief in the value of my work over the decades buoyed my spirits and helped me chart the seas of *Jump Time.* Jeremy is a magus to authors, a magician of ideas who sees a better future and then helps to create it. I am blessed in his trust and friendship.

Then there are the numerous students, especially those in the 1999 Mystery School who worked with this material in an expanded form along with many accompanying exercises and processes. As with all of my work, my close associate Peggy Rubin brought her special brilliance to helping develop and teach this material. Betty Rothenberger brought her depth of mind and reflection to each part as it unfolded.

And then there is my Airedale, Zeus Henry, who trained me in the nuances of new ways of jumping.

To the 1999 Mystery School,
whose participants embraced
and explored this material,
weekend after weekend and
chapter by chapter,
I dedicate the challenges
and ideas of Jump Time.

CONTENTS

THE SHOW MUST GO ON

It may be that some of you have opened this book because you are haunted by a specter: the grand finale of the world as we have known it. You know yourselves to be people of the parenthesis, living at the end of one era but not quite at the beginning of the new one. You may have labored in your various fields to make things better, and you may have tried to understand the change of which you are a part. And yet I suspect you to be frustrated, baffled even, by what must seem the most implausible, improbable series of happenings ever.

The maps no longer fit the territories. The only expected is the unexpected. Everything that was, isn't anymore, and everything that isn't is coming to be. Ours is an era of quantum change, the most radical deconstruction and reconstruction the world has seen. More and more history is happening faster and faster—faster than we can make sense of it. Life paths that have contained and sustained us across the millennia are vanishing as we speak, like the numbers of Gaia's species that are hourly becoming extinct. We are guests at a wake for a way of being that has been ours for hundreds, even thousands of years.

At the same time, we know that we are the ones who must go on. It is time for us to consider what is and what may be. Our agenda is nothing less than the future. Our challenge is to cultivate the vision and lay out the practical steps necessary to move through the opening times that follow upon closing times.

Unlike many others, you may be among those who refuse to believe that chaos leads to chaos, breakdown to catastrophe. You know that you have the power to direct the process along lines different from those that the prophets of gloom proclaim as inevitable. You know that the new millennium we have entered is the intersection between worlds, between species, between ourselves and forever. You know yourself to be its pilgrims and its parents. No old formulas or stopgap solutions will suit. For a new world to be born, we must bring a new mind to bear. Nothing less will do in Jump Time.

In a Jump Time like the present, we as a species stand at a crossroads faced with radical choices, any of which promise to make tomorrow look nothing like yesterday. Personal Jump Times, moments when our life path reaches a fork and everything afterward changes, are more familiar to us. How do we react? Do we sit at the end of the known and refuse to budge? Do we walk back up the path we have been following, hoping to return to familiar territory? Or do we follow the old show-business adage that, whatever happens, the show must go on?

I remember it plainly. I had been standing on the seat in the front of the train car singing a medley of "art" songs for the soldiers and sailors. I had just finished with an operatic rendition of "The Bluebird of Happiness" and "The Chocolate Soldier" and was moving into my bring-down-the-house number: "Ahhhhhh, sweet mystery of life, at last I've found thee . . ."

A Top Sergeant who had obviously been drinking interrupted. "Hey," he announced, "That ain't no five-year-old kid. That's a midget. No little kid's got a voice like that."

True, no little kid did—until this little kid came along. However it happened, by the time I was five years old, I had a ridiculously ripe mezzo-soprano singing voice.

My mother encouraged the phenomenon by giving me operatic singing lessons from the time I could carry a tune and by teaching me only melodies a diva might sing. Instead of "Three Blind Mice," I was taught to trill the passionate cadenzas of *Carmen*'s "Habanera." While other little girls were bouncing the ball and chanting "A, my name is Alice . . ." I was dutifully warbling "Mi Chiamano Mimí."

My voice and I were exhibited on every possible occasion—to visitors, cleaning ladies, door-to-door salesmen, the Chinese laundry man, on radio programs, in USO shows, and, of course, on the train as we crisscrossed the country.

While meteoric in its rise, my singing career was nevertheless short-lived. At the age of six, at the Hollywood Bowl, while singing for thousands of people, I lost my singing voice forever. Actually, I had lost it several days before that, but I never got the chance to explain that to the impresario or to my father, who set me up for the job.

The catastrophe began the day my father, a comedy writer with a good many show-business connections, came home and said to me, "Jeanie-pot! I've got a great opportunity for you. A couple of days after Christmas, you're going to follow Kate Smith at the Hollywood Bowl and sing 'God Bless America' after she finishes. OK by you?"

It certainly was. Kate Smith was my ideal, and so, though it was only October, I went into training like a prizefighter with a crack at a title bout. For nearly two months, I had but one theme, one song, one set of words, one response to every question.

"Jeanie, what would you like for dinner tonight?"

"Stand beside her, and guide her . . ."

"Jeanie, look at you. Your knees are all scraped and your hands and face are filthy. Where have you been?"

"From the mountains. To the valleys. To the oceans, white with foam . . ."

The rest of my repertoire abandoned, I pressed on, singing my one song—our song, Kate Smith's and mine—until I had acquired a kind of rare perfection, the ultimate in the singing of "God Bless America," a Rembrandt of song, a Praxiteles of performance. And then one morning in mid-December, while standing next to me in the spelling bee, Cookie Collozzi coughed in my face. Shortly thereafter I was coughing, and then I was whooping. I whooped and whooped and whooped out all of my high vocal chords. What was left was strictly in the basement.

My mother tried to assuage my bleats of grief by assuring me that the doctor had said that a few days of keeping as quiet as possible should bring back my normal voice. I tried to believe her, maintaining a strict silence throughout the train trip to Los Angeles, where my father would meet us.

"Jack, I have something to tell you," my mother began as he whisked us from the train, which had arrived several hours late, to a limousine hearse he had borrowed from a friendly funeral parlor.

"Later, Mary. We have exactly twenty-eight minutes to get Jeanie to the Hollywood Bowl."

With that, he slammed the back door of the ancient limo and jumped into the driver's seat. A glass partition kept him from hearing me croak about the state of my voice, and all our knocking and grimacing went ignored.

As we neared the Bowl, we could hear Kate Smith's magnificent contralto belting out the song over the amplifiers. The very hills shook in resonance, and there was no question that in this corner of southern California and for miles around, America knew that it was blessed. When the door flew open, my father dragged me out of the car, hoisted me to his shoulder, and started running toward the backstage area.

"Where have you been?" the stage manager shouted at us over the deafening applause, as Kate Smith took her bows. "We've been looking all over for you. OK, kid, you're next. Break a leg."

That sounded like a wonderful idea, but alas, there were no stairs in sight from which I might hurl myself. Instead, I was pushed on stage in time to hear the M.C. exclaim, "Now, I've got a special treat for you. Here, straight from New York City, to sing the same magnificent song, which has meant so much to so many during these times of war, is the little, little girl with the big, big voice—Jeanie Houston!"

Big applause. Thousands of friendly faces peered up at me,

freshly exhilarated by Miss Smith's rousing rendition. The orchestra played the introduction, and with no other options, I opened my mouth and began to sing.

Needless to say, my performance was greeted by a combination of catcalls, booing, and people getting up to leave. I sang on, however, pushing my gravelly basso through all the stanzas. At last, with a final growl of "My home sweet home," I zoomed off stage into the incredulous arms of my father.

"Why didn't you tell me you had lost your voice?" he demanded.

The show must go on. If ever there were a single edict, a primary directive for the human race, from whatever gods there are, this is it. Shakespeare and other seers into the nature of reality speak of the theater of the world, wherein the plot moves inexorably, carrying in its wake the rise and fall of souls and cities. Occasionally the plot quickens, and the play takes strange and deviant turns. Surreal surprises enter from the wings, the orchestra in the pit offers dissonance as well as sweet airs, and the gods, both in and out of the machines, offer more than we had bargained for. Such is the present drama that Mother Earth offers to her ticket-holders—every last one of us.

How often in these days do we feel that our lives are like that old dream in which we are on stage in a play, dressed in a period costume, with no idea what our lines are supposed to be? What used to be held as solid expectations—professions, relationships, religions—have become capricious; the old verities are lost, the anticipated outcome has vanished or been transformed into a mockery of itself. Our very identities seem to be shape-shifting. We open our mouths expecting melody, and a crow flies out.

Standing on the stage of your old certitudes, many of you, perhaps, have experienced such discomfiting surprises as these:

- The job or profession for which you trained and which you expected to follow for many years suddenly no longer exists.
- You fall love with an improbable person or idea, and in an eyeblink everything you knew or believed is gone. In the light of this tremendous circumstance, who you were yesterday bears little resemblance to who you are becoming.
- You wake up in the morning consumed by an urge to get on with it. What "it" is, you do not know, but it is barking at your heels like the Hound of Heaven. Something unknown is calling you, and you know you will cross continents, oceans—realities, even—to discover it.
- You are seized by concern about the disparities of the time, the widening gap between rich and poor, private affluence and public squalor. Moreover, you can't shake the conviction that you personally must do something to redress these wrongs.
- Words leap off the pages of books. Synchronicities abound. The universe is trying to tell you something, and you can no longer ignore its message.
- Your traditional religion no longer serves you, but one from an utterly foreign culture is speaking deeply to you.

Clearly you have come to the end of the road. An unknown future beckons. Way leads on to way, and only in retrospect do the turnings and returnings make any sense. Looking back on my aborted singing career, I see now that had I not lost the upper register of my vocal range, I would probably have ended up on the concert stage and never taken the fork that led to a very different though equally public sort of life. However, what goes around comes around, and my capacity for public mortification also moved to grander stages, the most recent being several years back when I ended up on the front page of most of the newspapers in the world as the so-called guru and trance medium of Hillary Clinton. To set the record straight, I did not download Eleanor Roosevelt from the cosmos in the solarium of the White House. I just helped Hillary conceptualize and then edit her book *It Takes a Village to Raise a Child.* The media, however, needed a more lucrative story, so they made up a character and a spooky plot and put my name to it. Again I lost the upper registers, not of my voice but of my reputation.

Nevertheless, the huge drama that is now unfolding on the world stage makes anyone's particular humblings look like so much hummus. The urgency behind my work, my sense of the importance of present history and its fierce momentum, helps me rise out of the flatland of my various levelings, croakings, and pratfalls and get on with it. In what follows I hope to persuade you to do the same—to stride boldly forward, even though where you might be heading and what you'll find there is not yet clear.

Perhaps you'll find, as I have, that life's jumps move you in unexpected directions. Instead of the theatre of the performing arts, my work has taken me into the theatre of the mind, the drama of the soul, and the Broadway of cultures as they recover and share their once and future genius. In conducting basic research into human capacities and applying its findings in human development programs over the past thirty years, I have seen that lost domains of consciousness can be recovered and put to practical use. When mind and body are reloomed, psychophysical abilities for healing and self-healing, for mastering an artistic or athletic skill, and for rapid learning are quickened.

I have seen that it is possible to think in images as well as words, to learn with one's whole body, and to harvest riches of creativity from interior landscapes. I have seen people discover the secrets of time—time in the body and time out of mind, in which one experiences subjectively a very short period of time as being much longer. In that short time, the mind can select, synthesize, and create in minutes what might normally take months and gain the same benefit from rehearsing an activity as if one had practiced for hours or weeks.

I have helped people play on the spectrum of consciousness so as not to be stuck in a single bandwidth and gain instead the option of selecting states of consciousness most conducive to concentration, mindfulness, optimal physical performance, creativity, or spiritual exercise. I have marveled that the world within is as vast as the world without and as filled with story, myth, symbol, archetype, god and goddess. These adventures of the soul convince me that we humans have the innate equipment to be actors, directors, playwrights, and producers on the world stage of Jump Time.

Beyond the laboratory and the seminar stage, part of my calling has been to study, collect, and apply human capacities as they have developed in cultures around the world. Years ago, the anthropologist Margaret Mead sent me out with letters of introduction to the elders of various cultures so that I might bring back some of the unique ways of being that ancient and indigenous societies have developed. From this beginning, my work for international development agencies gave me the chance to study firsthand how Africans walk and think and celebrate spirit; how the Chinese teach and study and paint; how Eskimos experience vivid three-dimensional inner imagery; how the Balinese learn to perform any manner of artistic endeavor so rapidly and with such high craft; how a tribe along the Amazon raises emotionally healthy children; why certain children in India raised amid traditional music develop extraordinary skill in mathematics. My work has taken me to many lands, where I treat cultures like persons and help them deepen and recover their genius, while they link with other cultures, including the emerging planetary civilization. And I treat people like cultures, helping them discover within themselves the rich strata of untapped potentials of mind and body.

For potentials are no longer limited by place and culture. In Jump Time's developing hybrid world, capacities once nurtured in separate societies are available to the entire family of humankind. This is a stupendous happening, as important as the discovery of new continents during the time of the great sea journeys. For the first time in human history the genius of the human race is available for all to harvest. These rediscovered capacities may be evolutionary accelerators, now being gathered from many places, times, and cultures to awaken our species to who we are and what we yet may be and do. Often, however, it is not comfortable. We can for a time find ourselves strangers in a very strange land, wishing we could return to the comforts of a more insular and familiar worldview. Yet when we get beyond the shutterings of our local cultural trance, we gain the courage to nurture the emerging forms of the possible human and the possible society.

With the present move toward planetization and with the entire spectrum of historical social development at hand, I believe that the world is set for the radical transformation that I call Jump Time. In this time of accelerating change, all cultures, regardless of their social or economic level, have something of supreme value to offer the whole. I have observed that too many members of European-derived cultures who revel in technique and objective mastery are sadly lacking in the spiritual awareness and subjective complexity found in aboriginal cultures belonging to earlier stages of historical development. True, the tragic intervention of colonialism dimmed these cultures for a while, but more recently, Maoris, Australian aborigines, and Native Americans, among others, are reviving their cultural wisdom. As the genius of many cultures is brought together, as I believe is now happening, the current crisis of social breakdown and moral disorder can be transformed into the creative symbiosis of the coming world civilization, of cultures seeding each other while preserving and enhancing their individual style and differences.

So what is Jump Time? It is the changing of the guard on every level, in which every "given" is quite literally up for grabs. It is the momentum behind the drama of the world, the breakdown and breakthrough of every old way of being, knowing, relating, governing, and believing. It shakes the foundations of all and everything. And it allows for another order of reality to come into time.

Jump Time is a whole system transition, a condition of interactive change that affects every aspect of life as we know it. The vision of change I am describing is generally optimistic. It focuses on the emergence of patterns of possibility never before available to the Earth's people as a whole. This optimism is, paradoxically, based on the recognition that virtually every known institution and way of being is currently in a state of deconstruction and breakdown. Given the scientific, technological, crosscultural, and social tools at hand, and given, too, that humanity is searching as never before to cooperate in so many areas, it seems feasible to me that we may be ready to integrate inner and outer dimensions of life in ways that infuse new depth into psychological and spiritual growth and new purpose and responsibility into social transformation.

As I see it, our current stage of collective growth is being propelled by five forces that, taken together, determine the direction of our jump into the future and where we may land:

1. ***The Evolutionary Pulse from Earth and Universe.*** Basic to Jump Time is my belief, widely shared, that we humans are not alone as we face the massive transition that is upon us. Rather, we are embedded in a larger ecology of being, its motive force arising simultaneously from the planet that is our birthplace and the stars that are our destination. Pulsed by Earth and Universe toward a new stage of growth, we are waking up to the realization that we can become partners in creation — stewards of the Earth's well-being and conscious participants in the cosmic epic of evolution. As ancient peoples have always known, the story is bigger than all of us yet desires our engagement, our love, and our commitment. People everywhere are feeling a call, a quickening, an energy that brooks no whiny naysaying, an invitation to the dance. Chapter 1 of this book, on individual and universal destiny, explores the evolutionary impulse that drives our growth and suggests how we might learn to ride and direct its energy.

2. ***The Repatterning of Human Nature.*** The second force of Jump Time pushes us to discover and utilize our dormant or little-used capacities and to come to a more comprehensive understanding of our place and responsibilities in this world and

time. In Jump Time, I am convinced, capacities that have belonged to the few must become the province and requirement of the many if we are to survive the next hundred years. We must each learn to tap into the creative workshops of the mind to solve problems and to bring forth art, poetry, invention. We must discover ways to feel at home with anyone, anywhere, at any time. Most people, given opportunity and training, can learn to think, feel, and know in new ways; function in their bodies with better use and awareness; become more creative, more imaginative, and aspire within realistic limits to a much larger awareness, one that is better equipped to deal with the complex challenges of life. The consciousness that solves a problem cannot be the same consciousness that created it. But the consciousness that can rise to this occasion needs models of its own matured possibilities, visions of what the possible human can be and do that go beyond the limitations of academic excellence or dogged persistence to attain certain goals. Chapter 2, on the psychology of the Jump Time self, and Chapter 3, on educating our children and ourselves for the future, look at what might inspire this repatterning.

*3. **The Regenesis of Society.*** As the self is repatterned, the ways we relate to each other are necessarily shifting as well, toward the discovery of new styles of interpersonal connection and new ways of being in community, given a global society. The movement seems to be from the egocentric and the ethnocentric to the worldcentric—a fundamental change in the nature of civilization, compelling a passage beyond the mindset and institutions of millennia. Critical to this reformation is a true partnership society, in which women join men in the full social agenda. Since women tend to emphasize process over product, being rather than doing, deepening rather than end-goaling, it is inevitable that as a result of this partnership, linear, sequential solution-seeking will yield to the knowing that comes from seeing things in whole constellations rather than as discrete facts. The consciousness engendered by this comprehensive vision is well adapted to orchestrating the multiple variables and getting along with the multicultural realities of the modern world. It raises hope for forgiveness between and healing among nations and ethnic groups. Essential to this matured consciousness is moral and ethical growth toward an empathy between individuals and nations that honors the golden rule of human interchange.

The regenesis of social forms also asks that governments no longer engage in social engineering to fix specific problems but rather that they understand the world as an ecology, a complex adaptive system in which global awareness is applied to local concerns. Here we need models of a new order of relationships and their place in a possible society, one in which male and female, science and spirituality, economics and ecology, civic participation and personal growth come together in an integral and interdependent matrix for the benefit of all. Chapter 4, on new ways of seeing our interpersonal relationships, and Chapter 5, on a revisioning of the relationships between peoples and governments, explore this agenda.

4. The Breakdown of the Membrane. In the interdependent world of Jump Time, old barriers are dissolving along with the phobias that sustained them. The new technologies of instant communication and exchange of information enable people to join minds and hearts in mutual discovery and creation. In the Internetted world, more and more people are coming to accept the benefits of cultural diversity and to adopt a more inclusive worldview, inspiring human beings with renewed hope and caring. What began in migrations and global economics is fast becoming a worldwide network of individuals and institutions quickened by the desire to create a new social paradigm, in which humanity and the Earth are each enhanced within the context of a collective destiny. As the membrane of old forms breaks down, a more complex and inclusive global organism comes into being. As living cells within this new organism, we are rescaled to earth-wide proportions in our responsiveness—and our responsibilities. Chapter 6, on the developing fusionary world culture, and Chapter 7, on the Internet and its effect on the world mind, address these issues.

5. ***The Breakthrough of the Depths.*** The depths are breaking through most apparently in the spiritual renaissance that is occurring everywhere in Jump Time. Not since the days of Plato and Buddha and Confucius, some twenty-five hundred years ago, has there been such an uprising of spiritual yearning. And what with the inevitable crossfertilization of the wisdom and practices of world spiritual traditions, more and more people are gaining access to the Source of our being and becoming. As a new story begins, the mythopoetic understandings of many cultures rise and converge. Archetypal ideas and symbols spring into consciousness or are consciously sought in the popular culture. Either the depths are rising because the time requires it, or we are being impelled by the times to explore those depths. However it is occurring, our spiritual preoccupations are essential, given that we need a spiritual renaissance commensurate with our technological developments to give us a sufficiency of inner inspiration to guide our expanding outer forms. Chapter 8, on the spirituality of Jump Time, reflects on this phenomenon.

All books involve a certain kind of conversation between author and reader. I am inviting you to go further and enter into an imaginal space where you will join me and the others reading this book in jumping onto the next path. I offer to be your guide to our present Jump Time and to the exploration of previous times of world transition that have much to teach us about how to cope creatively with what we face. For the world has known other Jump Times, though never so consciously or with so much to gain or to lose.

The Earth is a hothouse now. Six billion members of the human family and rising, congregated together on a spinning ball, in stress, in ferment, caught between what was and what is yet to be.

It is time to ask the great questions: How can we make a better world? What must we do to serve the larger story? These questions help us clarify and define our objectives. They prompt us to articulate goals lofty enough to lift us out of petty preoccupations and unite us in pursuit of objectives worthy of our best efforts.

The world is hungry for vision. At a time when whole systems are in transition and global forces challenge all authority, there is an insistence in the mud, contractions shiver through the Earth womb, patterns of possibility strain to emerge from the rough clay of changing social structures.

It is a matter of kairos, the potent time for fortuitous happenings. In ancient Greek, *kairos* referred to that moment when the shuttle passes through the openings in the warp and woof threads. In the loaded time when such things happen, the new fabric can take form.

Right now is the right time to make things right. I invite you to join me on the stage of forever, where what we do may raise the curtain on the world's next act.

CHAPTER 1

ARCHIMEDES AND THE EVOLUTIONARY LEVER

My parents came from vastly different backgrounds — my father a Texas-bred comedy writer and chiropractor, my mother a Sicilian economist and method actress. Though my folks tried valiantly to bring their divided and distinguished worlds together, they never quite made it. They divorced when I was fourteen.

I see myself back then, a child of nine or ten, trying gallantly to interpret my parents to each other.

"Dad, the smell of frying garlic isn't so bad. Your Uncle Johnny's hog trough in Beauford, Texas, stinks much worse."

"Mother, I know you'd rather listen to the Metropolitan Opera on Saturday afternoon, but just once couldn't you try the Grand Ole Opry? Dad loves it, and the singers are just as loud."

Perhaps I've tried to make up for my failure with my parents, for when I grew up, my work centered around bringing cultures together, sometimes the mindsets of our many states of consciousness and frequently the traditions of far-flung regions of the Earth. I've discovered that in the meeting and melding of dissimilar ways of being, strange and wonderful things happen.

As was not the case with my parents, whose tribal boundaries never permanently relaxed, exchanges between the world's peoples—Bangladeshi, Maori, Balinese, Burmese—are frequently dynamic and fruitful. Over the past twenty years, I have spent part of each year working in countries around the world, often at the invitation of international development agencies, helping people find ways to maintain and deepen their unique traditions while they discover their part in the emerging world story.

This work has given me a finger on the pulse of what is happening. Increasingly I sense a new cultural music coming into time. A welter of multiple meters and offbeat phrasing, it is a coding of creativity and imagination, a counterpoint of styles of knowing and being. As cultures come together and exchange their essence, they join in a cadence of awakening, a new idiom of consciousness that is exhilarating and revolutionary. This rhythm carries us from the ballads of local concerns to the concerto of a larger ecology of being. Within it we feel *the evolutionary pulse arising from Earth and Universe* and come to understand that our individual life is part of the unfinished symphony of the cosmos. A strange, sweet music is coursing through us. Each of its notes carries with it a task, a responsibility: play your part, it hums to us, for what you do profoundly makes a difference.

Whence does this evolutionary pulse arise? From God? From hyperdimensional realities? From the Big Bang? Could it be that something in our flesh remembers the dawn of the universe, in fact, partakes in it, for aren't all particles one particle? On some sonic register beyond the threshold of hearing, perhaps, our ears still vibrate from this sound that contains all sounds, carrying the message that everything is energy, vibration, frequency, resonance. Even the most solid object is ultimately a dance of changing energy patterns. Ultimately all is rhythm, all music. The world is sound. *Aum,* the beginning, heard now and forever. This primal sound pulses us as it gives rise to the complex webbing of interdependent relationships through which our own lives are embedded within Larger Life.

In the century now past, when cultures with conflicting priorities and values came into contact, their meetings all too often exploded in traumatically passionate ways—world wars, iron curtains, nuclear weapons, clashes in which each side saw the other as alien. Better by far, now that we know better, is this new mitotic exchange of cultural DNA, the strands of many times and traditions carrying genomes of familiarity and fascination instead of insularities and phobias. With the breakdown of walls between people and nations that technology has wrought, we are witness in our lifetime to a massive ontological shift in personal and social consciousness.

We sense that we have crossed the threshold into another reality, almost another planet. Once distant seas of consciousness crash together and churn the evolution of the world mind—a new Jump Time mind now coming into being.

The energetics of cultural evolution were brought home to me in a most literal and powerful way in September 1992, when I was accompanied by several Maori elders to the northernmost point of New Zealand, or Aotearoa, as the Maoris call their beautiful land of the long white cloud.

We stood on a ridge at the end of the island looking down to a place where three seas, the China Sea, the Tasmanian Sea, and the Pacific Ocean, crash together in roiling waves. Each of the seas brings its own winds, so one is buffeted by gusts from all sides. Let me invite you to stand there with me.

Something is uncanny here. There is Presence here. Could it be that the waves also bring with them the spirits of other lands? In one way or another, these seas have touched upon all seas and every shore. Are they bringing global news of nature's plans and human folly?

The wind is growing stronger. The waves collide in some giant argument, or is it a mating, an exchange of essences? Is this the place where the planetary DNA is coded anew?

"It could be," my hostess shouts over the roaring of the winds. "But it is also the place where Maoris go when they die to lift off to the Other World." She points to a small spit of land where the sea currents come together in a boiling vortex that seems a plausible place to leave the planet.

My hostess is a wise woman of the Maori clan. Her son, with whom I had spent the previous day, is himself a receiver of the waves of human speculation from many lands. A holder and practitioner of traditional Maori wisdom, he has a physicist's knowledge of the new science and is a speculative thinker of the first order. He is also a storekeeper.

I have found that the Maoris, perhaps more than any other indigenous culture I have worked with, have made considerable strides towards restoring their culture, even after a hundred and seventy years of colonial rule. In spite of the problems that afflict all indigenous peoples, they are relearning their language, winning back their land and fishing rights, rebuilding their communities, developing their ancient arts and sciences, and generally moving as a culture toward the fulfillment of their innate potential.

Many among them, like my hostess and her son, have acquired, appreciated, and even improved upon the knowledge of other cultures. The previous evening, the son astounded me by explaining how the ancient principles of Maori metaphysics anticipate those of quantum physics. I ask my hostess at the top of my voice to tell me how it is that the Maoris have recovered so much of their genius.

"It is because of places like this," she bellows, "where the spirits of many peoples and many lands can meet and refresh themselves. Here, as well as in our great meeting houses, our Maraes, we remember who we are . . ."

Her voice takes on new power, equal to the winds. *"And call our spirits home!"*

Here and now, in a new millennium, we too stand on the shore of time, at the edge of history, receiving the winds of change. We welcome the homecoming of the spiritual force that can quicken our passion for life in the Jump Time.

Fears abound. Few, if any, have been trained for such a time. We feel cluttered and burdened with the learnings of a world now passing away, shackled by beliefs birthed out of a narrower view of the cosmos. No matter how "postmodern" we pretend to be, each of us has been marinated in the medieval soup of the mind. To face the radical needs of the future, we need a new natural philosophy, one that encompasses an appreciation of what is emerging in science and psychology as our evolutionary possibilities.

I often attend conferences of high-minded folk from many disciplines who address the problems of an old world passing into a new. The millennium has brought many to the table. Some are radical visionaries proposing utopian global house cleanings. Others are reformers devoted to redressing old wrongs. Each attempts to push the membrane that wraps us in unknowing. With all their practical skill and accomplishments, they are humble before the mystery of a world in transition.

"It is as if we are in a giant womb, trying to figure out what happens next," a United Nations official said to me recently at the State of the World Forum.

I was startled by her remark, for earlier that day, I had been chatting about international politics with a Bulgarian cab driver as he drove me around San Francisco at dawn looking for an open Kinko's so I could print out from disk the speech I was to give.

"I have just been present at the birth of my daughter," he told me. "I was very afraid, for I had never seen such a thing. It was very messy and very beautiful. And after all the hours of my wife's labor and the painful contractions, a new life! Maybe that is what is trying to happen in our world."

I often describe my work as a kind of midwifery. Organizations and cultures, as well as individuals, sometimes need a steadying hand as they birth themselves into a world as strange and unexpected as the one babies face when they emerge from the womb. If we are to grow up to be stewards of the Jump Time that is upon us, I tell them, we must make the best use of our capacities.

Our bodies and minds are coded with an extraordinary array of possibilities and potentials. Some no doubt were cooked in ancient caves. Others were laid down even earlier. On the genetic scale, we humans encompass all that has gone before. We are inheritors of the star stuff from which life came and relatives of every organism on the planet. The remembrance of things past is coded into our bodies. We crawl with single-cell organisms and are home to a universe of bacteria. Briny oceans pool in our blood. Our brain is a nested creature, its oldest reptilian and amphibian core covered over by mammal and human brains—and the promise of something burgeoning, something more.

When my research has given me the opportunity to take depth soundings of the continents within human beings, I have marveled at the enormity we contain. Somewhere in the vast treasure trove of the body/mind, I am convinced, we remember everything we have learned over the past fifteen billion years. We are the ultimate evolutionary hybrids, and the vigor of human genetic inheritance, if we could but claim it and work with it, is more than enough for us to get on with it.

Michael Murphy in *The Future of the Body* argues persuasively that human capacities stand on the shoulders of earlier evolutionary developments, even as they move toward extraordinary complexity and application.

> We can cultivate somatic awareness and control . . . because nerve cells that evolved from analogous structures in the earliest vertebrates are deployed throughout our bodies. Relaxation exercises are effective because we possess a parasympathetic system that developed during the long course of mammalian evolution. We can become creatively absorbed in work, perhaps, because we have inherited capacities for catalepsy, analgesia, and selective amnesia that facilitate escape and hunting.

> In short, self-regulation skills, regenerative relaxation, and performance trance, like other kinds of creative functioning, are based on capacities that developed among our animal forbears. And while transformative practices draw upon our animal inheritance, they also employ uniquely human activities. The imagination we use to enjoy books can be cultivated to induce metanormal cognitions or to facilitate extraordinary physical skills. The self-reflection we sometimes practice when confronted by difficulty can be deepened by means of sustained meditation. (Los Angeles: Tarcher, 1992, p. 543)

Evolutionary Jumps

Each evolutionary jump that we can name in the history of the universe is clear to us because some human jumped to see it and then created a story about it which we could understand. A primitive hand finds a stick, sees it as a tool, and uses it to find food. Guttural sounds deep in the throat give warnings, directions, and eventually express thoughts and feelings. Ice melts; caves give way to huts and houses; women collect wild seeds and poke them into the ground. A man watches grapes being crushed, and the printing press is born. Moon longings become moon landings. What we see in evolution is the push toward greater levels of complexity—increasing diversity, organization, and connectivity. Looking back at how we got here may provide us with clues to where we are headed.

The first jump was the Big Bang, an explosion of light and sound so intense that whatever was before or from some otherwhere imploded itself into a tiny charge of hydrogen and jumped into universal form.

Then, some five billion years ago, a supernova reached Jump

Time and, with an unimaginably fierce explosion, offered itself to the Universe in billions of pieces.

Another jump came for us when elements spun out of the explosion, coalesced into a ball that condensed into our mother, our molten planet.

Many jumps of cooling, crusting, boiling, steaming, raining. Seas form, the crust roils, land masses jump and shift, break off and crawl over each other, pulverizing everything in their wake. Meteors whiz by, raising blinding dust storms, or carom into the Earth's crust, tossing pieces of it sky-high.

It's a setup for the biggest jump of all—life. Lightning spikes the Earth. Molecules break apart and recombine, jumping through change after chemical change.

At the edges of the rock in the shallow waters, the feast of life begins. Giant proteins play with RNA and DNA. Some jump to form enzymes, which hurry everything along. Nucleic acids hold the information to make the huge jump to self-replication. Molecules build on each other, combine and recombine, building ever larger structures, modeling and learning from each other.

JUMP! Cell walls that move and flex, letting food in and wastes out.

JUMP! Bacteria form, creating food for themselves. As a waste product—a deadly poison, oxygen, is released.

JUMP! Cope with oxygen.

JUMP! Grow or die! Cooperate or perish! Make agreements! Enjoy diversity. Put it to work.

JUMP! The mothers and fathers of us all are born—the nucleated cell!

JUMP! Cell mitosis.

JUMP! Sexual reproduction—organisms unite so the species can continue.

Plants, JUMP! Sea creatures, JUMP! Land creatures, BIG JUMP! Dinosaurs, insects, flowers, birds.

JUMP! Heads up, snouts rise, eyes converge, brain grows!

JUMP! Walking upright.

JUMP! Vocal communication.

JUMP! Awareness of ourselves.

JUMP! Tool making, clothes wearing, plant gathering, seed setting, wheel turning, horse taming, water channeling, plow furrowing, loom spinning, food storing, star charting, tower building, metal smelting, myth telling, hieroglyphs and cuneiform and alphabets to record all this.

JUMP! Bureaucracy and empire, religions and scriptures. The Buddha, Confucius, Pythagoras, Christ.

JUMP! Book scribing, print pressing, art making, play writing. Telescopes, microscopes, spectroscopes, stethoscopes. Renaissance, revolution, resettling, migration. Liberty, fraternity, equality, democracy.

JUMP! Vaccination, sanitation, medication, refrigeration. Computers! Air travel, space travel, life extending, gender choosing, species making. World links, world banks, global village. And here we are, because it's JUMP TIME!

And the jumps continue, for it seems inevitable that future developments will follow this path. Everything is accelerating, the jumps coming closer together. Human knowledge is doubling every ten years. The extent of our mapping of the genetic sequences of the DNA molecule doubles every two years. Computer power is doubling every eighteen months. The Internet is doubling every year. And of course, almost daily, we read of new advances in space exploration, computer technology, medical science, and telecommunications, each jump in complexity occurring in a fraction of the time of the previous one. Reading the science pages of *The New York Times,* I am reminded of the words of the prophet Daniel: "Many shall run . . . to and fro, and knowledge shall be increased" (Dan. 12.4).

Churning as we are in this sea of so much change, jumps abound in our everyday lives as well. Our lives now regularly contain many times the amount of experience of our ancestors of earlier centuries; as a result, our personal jumps seem to be accelerating in frequency as well as amplitude. We are heirs to an extraordinary speeding up of the evolutionary process. We jump to new professions, partners, lifestyles, and religions seemingly at will. Nothing, it seems, is impossible for us. Nature, through us, seems to be entering a new epoch—not so much *biological* evolution but *conscious* evolution. We have become conscious of our capacity to direct the next phase not only of our personal lives but of the world's destiny as well.

Lessons from Evolutionary History

To understand what we are facing, we might consider the evolutionary journeys of those simpler organisms who are our precursors. University of Massachusetts microbiologist Lynn Margulis suggests that we reexamine our kinship to the teeming world of the very tiny. Bacteria, she points out, are the ur–life form, microbial architects, essential to all living things. Without bacteria, life and its processes would cease.

In the first two billion years of life on Earth, bacteria continuously transformed the planet's surface and atmosphere, inventing life's essential chemical systems—fermentation, photosynthesis, and oxygen breathing and fixing atmospheric nitrogen into proteins. They too faced crises of population expansion, starvation, and pollution. They survived these challenges because they developed remarkable "evolutionary organs"—capacities that allowed them to share and transfer genetic information.

Contrary to the view that evolution is a combative, red-in-tooth-and-claw struggle, bacteria led the way by networking. Microbial life forms multiplied and grew more complex by co-opting others, not just by killing them. They could also merge—combine their bodies, form permanent alliances. Symbiogenesis, the merging of organisms into new collectives, evolved as a major strategy for environmental survival.

The parallels between ourselves and the world of the microbes—our ancestors and progenitors—are both humbling and hopeful. The lesson of the bacteria is that only through interaction—solutions to problems drawn from the collective of which we are all part—can we grow to the next stage.

Paleobiology offers another provocative evolutionary parallel to Jump Time in the notion of punctuated equilibrium, or "punk eek," as it is called in scientific circles. Change, evolutionary theorists tell us, doesn't happen gradually. Rather, things go along as they have been going for a long while—in a state of equilibrium—until a species, living at the edge of its tolerance, experiences enough ferment and stress to punctuate the equilibrium with a sudden jump to a whole new order of being.

The theory was proposed in 1972 by Stephen Jay Gould of Harvard University and Niles Eldredge of the American Museum of Natural History. They challenged the traditional Darwinian view, known as phyletic gradualism, which held that changes in any species occur over long stretches of time.

As Charles Darwin himself pointed out, great gaps in the fossil records make it impossible to trace conclusively the emergence of new species. He hoped that those gaps would eventually be filled in by new fossil discoveries such that natural selection could account satisfactorily for the gradual transformation of one species into another.

Gould and Eldredge noted, however, that those troublesome gaps in the fossil record were not being filled in, though more than a century had elapsed since publication of Darwin's *Origin of Species*. In fact, as the fossil record became more complete, it seemed clear that species persisted in sameness for time out of mind before they suddenly gave rise to new ones in what amounts to the geological blink of an eye. Eons of sameness were punctuated by the abrupt emergence of new evolutionary forms within relatively few generations. Evolution, Gould and Eldredge concluded, is not always a process of steady and gradual change but sometimes involves species "jumping" from one state of being to another.

The question remains, why does a species punctuate its nice long equilibrium snooze? Does its environment change? Is it living at the edge of its tolerance? Is it bored with itself? Is there a great plan or pattern coded within it that responds to a stimulus and initiates transformation after a certain period of time?

When it comes to the fossil records, clues are widespread and suggestive. In what is known as the Judith River formation of Montana, for example, fossil evidence documents five million years of evolutionary stasis for a number of dinosaur groups. Then a rise in sea level put the Judith River area under water for some five hundred thousand years. A portion of the dinosaur population retreated from the flood to a smaller, more limited Montana habitat. They faced the stress of the new environment—Jurassic Park it wasn't—and they were isolated from the homogenizing effects of the larger dinosaur population. Within half a million years, no time at all on the evolutionary calendar, several new species evolved. When sea level fell again, these new dinosaur types moved back to the original dinosaurs' stomping grounds, spreading rapidly through the Judith River area.

When it comes to the human record, we have only to look around us to see ample evidence for cultural jumps as comprehensive and as vivid as any the dinosaurs experienced. Take something as simple as accounting. A hundred and fifty short years ago, say, in the England of Charles Dickens, clerks sat on high stools, their inky hands pushing scratchy nibs across huge ledgers as they kept their spidery accounts. Jump to today and data-entry clerks are keying in numbers on spreadsheet programs that perform complicated calculations and graph years of transactions in microseconds. From the abacus to the ledger book is not a great jump—just a few way stations of records on clay tablets and papyrus. But from the ledger book to the spreadsheet is a jump so huge it boggles comprehension.

What is looming before us now is a collective jump—faster and more complex than any the world has known. We find ourselves at present in the midst of the most massive shift of perspective humankind has ever known. Clearly we are living in a time in which our very nature is in transition. The scope of change is calling forth patterns and potentials in the human brain/mind system that as far as we know were never needed before. Knowings that were relegated to the unconscious are becoming conscious. Experiences that belonged to extraordinary reality are become ordinary. With the intersection of so many ways of being from all over the planet, the maps of the psyche and of human possibilities are undergoing awesome change.

Aldous Huxley described the scope of our contemporary dilemma in his last utopian novel, *Island* (New York: Harper and Row, 1962, p. 134):

> Science is not enough, religion is not enough, art is not enough, politics and economics are not enough, nor is love, nor is duty, nor is action however disinterested, nor, however sublime, is contemplation. Nothing short of everything, will really do.

"Everything" is a tall order.

We humans have always been convinced that we can be more than we are, and many of us have suspected that everything is within reach. Scripture and folk tale abound with stories of people who have gone to their edges—and discovered that they could go further. Whether in feats of extraordinary skill or strength, bold quests into science or art, or explorations into the geography of inner or outer space, the human species is unique in its need to go "where no one has gone before." The varied expressions through history of our need to understand and accomplish everything arises from an evolutionary impulse that is biologically, psychologically, and spiritually innate.

Whatever healing fictions we tell ourselves to explain this evolutionary programming—that we were created in the image of an all-knowing God, that we contain within us the capacity for "waking up" to Buddha-like compassion and wisdom, that human development has been seeded and guided by extraplanetary beings who wish to help us assume our rightful place as galactic cocreators—we know at our core that our potential is limitless.

And yet, sometime during the century now dawning, we also have the power to abort the entire enterprise. Is our lovely blue-green planet to be a fatal mishap on a wing of the Milky Way, or a fifteen-billion-year project now finally coming to term? That question lies at the heart of Jump Time—the time to grow or die. In whimsical moments I wonder if the UFOs people keep seeing bopping around the planet may actually be full of spectators at the biggest sporting event this side of the galaxy, ETs on the edge of their seats betting on whether we will make it or not.

In a Jump Time with everything in transition, we can no longer afford to live as remedial members of the human race. A new set of values—holistic, syncretic, relationship and process oriented, organic, spiritual—is rising within us and around us. Though the forces of entropy and fear seek to contain or regress us, we know there is no going back. Our complex time requires a wiser use of our capacities, a richer music from the instrument we have been given. The world will thrive only if we can grow. The possible society will become a reality only if we learn to be the possible humans we are capable of being.

A Renaissance Jump Time

What must we do to become stewards of our own evolution? What must we do to move ourselves and the world to the next stage?

To answer this, we might try an experiment in going back to the future, specifically, to Renaissance Italy in the fifteenth century, a Jump Time not unlike our own. The Renaissance was literally a new birth—a breech birth of the human soul. In the midst of vast changes and upheavals, the artists, philosophers, and explorers of Florence entered the future facing backward toward the past. They dug in the earth, and they excavated in monastic libraries. They retrieved ancient genius from before the fall of Byzantium and found there the ideas and images of the Greeks, the Romans, and the Hebrews. The lost legacy of the world's past thoughts and dreams was born again into their time, and from this stimulus they grew a new body and a new mind.

Their imaginations were fired to dream again, but this time another order of reality. With the help of philosopher/mages like Marsilio Ficino and Pico della Mirandola, they discovered their inner world as an infinite landscape, as varied and companionable as the earth without. And since this inner world lay close to the Mind of the Maker, they were flooded with visions of what could be, pouring their inspiration forth as painters, sculptors, musicians, architects, engineers, scientists. Some, like Leonardo da Vinci, were all these at once.

It is my belief that the world's present Jump Time is cousin to this earlier one, not only in the bounteous palette of ancient images available to us but in the canvas of all times and places on which to display them. Today, the colors of world culture, the soul of the planet, can be reached through the tap of a computer key.

However, in Renaissance Italy the scales were in balance. Its thinkers and doers reveled in a sufficiency of stimulus without inundation. Not for them the ocean floods of muchness that threaten to drown us today. They had time to observe and digest and, like Leonardo, to ask Nature about herself, to observe the flight of birds and the fall of water, to study faces until they became their own. They visioned the geometries of invention on the whitewashed walls of the mind, linking inner and outer worlds, and built their palazzos there before they brought them into form. They studied everything that interested them and expressed it in many forms. They sculpted and painted, wrote poetry to extol what they loved, and captured its sounds on the lute. Their passions and their expression were keen and vital. No chisels carved away their ecstatic edges.

In their recovery of the past, they dialogued with ancient men and looked to them for inspiration. One of Leonardo's favorites was Archimedes, the great geometer and mathematician of the third century B.C. So intense was Archimedes' passion for his speculative work that he would forget to bathe. When his servants could not stand the stench any longer, they would pick him up and carry him against his will to the public baths. And yet, even there, he would trace out geometrical figures in the ashes of the chimney and draw lines with his fingers on his naked body while they were anointing him with oils and sweet savors. His delight in the study of geometry took him from himself and brought him into ecstasy.

Such passion comes from a confluence of desire: the inclination of an individual joined with a universal necessity whose season of completion has come. The pulleys and levers Archimedes drew linked him to archetypal patterns, the very mind stuff of the cosmos. Because his work connected him to cosmic knowledge, Archimedes shared in the universe's enthusiasm for elegant form and practicable function. Like Archimedes, Leonardo was happiest when the solution he devised to a problem helped to reveal Nature's laws. Standing in the meeting place between personal will and universal need, Archimedes' sense of possibility was limitless: "Give me a lever long enough and a place to stand," he said, "and I will move the Earth."

This metaphor and its message is particularly relevant for us today. It seems to me that we each must do just what Archimedes said: stand in the place of our own truth and hold the lever that is the highest expression of our individual destiny. The philosopher Aristotle had a wonderful word for destiny. He called it the *entelechy,* the dynamic purpose that propels us toward fulfilling our reason for being alive. When we wield the lever of our highest purpose, our destiny links to the unfolding of the universe. Our perspective shifts, and the Earth moves. The flat plane of our limited understanding becomes fully rounded, fleshed out in all dimensions, and our human soul gains the perspective to view itself.

In times of renaissance, when the landscape of future history stands vast and open before us, we have a rare opportunity to put our imagination to work reinventing ourselves and our civilization. With the lever of the entelechy, we can move the Earth beyond the eclipse that has kept us too long in the dark.

But as we move into the future, a cosmic humanism must enlighten our actions; our evolutionary jumps must be informed by the knowledge that we are acting on a global stage. As in Leonardo's drawing of the universal man, we are inscribed in harmony within the square of our immediate time and place but also contained within the circle of our infinite relations. In inspiration or ecstatic trance or as a result of long hours of study and search, we surrender to our larger nature. Then content arises from the inner world and, deeper still, from the Mind of the Maker. Michelangelo imaged what happens next in a panel on the ceiling of the Sistine Chapel. God and Adam reach out their hands toward each other. In the tension of the separation, a contact point is bridged, energy jumps across the arc, humanity becomes inspirited. This linking is what life is about, each of us called forth and connected to the universe's transcendent purpose.

Yet for us something is missing. Though we are fast becoming creators ourselves—our engineers reversing the course of rivers, our biotechnicians designing new species of plants and animals—we seem to have lost our moral compass; we follow the seduction of invention without responsibility. Even the angel of human creativity can breed monsters and grotesques without moral purpose to guide it. Jump Time asks us to cross a bridge, to stretch out hands and minds and hearts to be met by a destiny that is the world's highest as well as our own.

Finding Our Own Evolutionary Lever

How do we find the evolutionary lever within ourselves to lift all that we are to the next stage, so that can see deeply, act with power, be adequate stewards of Jump Time?

Too often, transforming ourselves seems to require a miracle. But miracles are merely the conscious activation of more patterns of reality than are usually seen by ordinary consciousness. Sometimes all that is needed is a shift in perspective. Leonardo's experiments in paint and his intricate scientific sketches had far-reaching consequences for the Western mind. When two-dimensional diagrams could depict conceptual detail with accurate perspective, flying machines could be built, telescopes and microscopes could be devised, architects could construct churches with soaring unsupported domes, explorers could journey beyond the planes of the set horizon to discover continents, infinities could be captured in formula if not in form, and we could extend the empire of humankind over things.

In spite of all that was gained, what was lost in Renaissance humanism was the comfort of the medieval notion of a God-ordained hierarchy in which everything existed in relationship to everything else—the social order reflecting the natural, cosmological, and celestial hierarchies, all held together in a Great Chain of Being whose figures artfully mirrored each other. Come the Renaissance and the shifting of perspectives, and, as John Donne lamented, "'Tis all in pieces, all coherence gone."

Not only was the world without fractured, but the many parts of the self that hitherto had been seen as sacred and as participating in the invisible purpose and meaning behind things were split off and sent reeling. Stripped of our cosmic connection, we were left falling into a black hole: "The sun is lost, and no man's wit knows where to look for it." Was it any wonder that an ego-centered psychology, affirming the importance of the individual over and against the world, and an increasing emphasis on economics and material consumption came to dominate human affairs? Too often and in too many places, reality was diminished to the flattening of one's spirit and the expansion of one's purse.

Now, as this era and its excesses comes to its end, another perspective looms, one that reaches back toward the soul-centered, nested universe we thought we had lost. People everywhere are regaining the sense that what they do matters profoundly to the course of events. Individual purpose and the world's greater destiny are renewing their connections. The images we hold, the thoughts we entertain, are reweaving a worldwide web of kinship.

When individuals come into resonance with universal purpose, they know it in their hearts, they feel it in their bones. There is a great assent, a cosmic yes, an arc of energy across the void. What stands revealed in such moments is the entelechy, the creative seed of greatness each of us contains. Some people are given very young to an innate sense of their essential reason for being—the oak their acorn is destined to become. For others, even adults living full and successful lives, the issue is activating an awareness of what more is possible. Many people I know, despite manifold professional accomplishments, are still wondering what they will be when they grow up. Few realize the answer completely. But when they do, their names become scriptural, for entelechy is the matrix of forms, the resonance of the divine in the human. It is ourselves writ large, the cosmic persona tuned to human purpose and possibility.

Contact entelechy, and all circuits are "go." Tune to it, and another order of perspective is at hand, one that comprehends the spatial and the temporal, that lifts the Earth of one's seeing into another domain where love rules and the patterns of higher governance are known. Words cannot really describe it. Metaphors fritter and fry in the fires of analogy. Entelechy is known in its experience. It is being in the flow. It is cooking on more burners. It is making the highest use of skills one has acquired. It is putting old capacities to work in new ways and discovering capabilities we never knew we had. It is growing the evolutionary organs of our future, transcendent selves.

When we live in service to our entelechy, we comprehend the genius of Leonardo, the compassion of the Buddha, the social consciousness of Martin Luther King, Jr., the word craft of Emily Dickinson. We become actors on the stage of a new story, our personal play a scene in the sacred drama of all times and places. We experience profound joy, a sense of blissful felicity. We enter the domains of the mythic and come face to face with the fullness of what we are.

What is it that holds us back from being all that we can be? What is it that keeps us blinkered and blinded, cycling round and round in the same well-worn tracks of conception and action?

While considering this, I look out my window and see the rocks and boulders that abound on my acres and, indeed, in my county. My home is in Rockland County, well named because a long time ago, during a major ice age, a glacier swept through this region carrying the detritus of mountains crumbling in its path. The glacier moved on and melted, but the rocks remain, mute testimony to the fierce ravages of nature.

Our personal wounds can also feel like the ravages of an indifferent Nature moving cruelly to its own purposes, running roughshod over our lives, chilling our hopes and dreams. The icy chemistry of fear, the glacial stupor of habituated thought, the Artics of ancient grievances all too often hold us frozen in structures that preclude growth and change. Emotional boulders choke our landscape, and we are caught in the trenches of unhappy memory.

Lift our gaze to the rest of the world, and the story seems much the same—people and nations frozen in postures of pain and parody, the atavisms of archaic hatreds, and the chill that precludes the healing of nations. What we need is the lure—the large enough vision or idea that can call us individually and collectively to the next evolutionary stage. If we could but hear it, the lure is already there, heard in the thunder of change that is the undertone of Jump Time.

Jump Time challenges us to melt our frozen winter consciousness in the upstart spring of a bright future for ourselves and our world. Almost every spring, when the ground thaws, my husband Bob and I prepare the soil and plant the seeds for our vegetable and herb garden. And every spring, Rockland County seems to heave up a fresh outcropping of rocks to be levered out of the ground and hauled away. Some of these rocks we turn into walls to keep out the deer, rabbits, and hedgehogs, already gleeful in expectation of the bounty our garden clearly means for them. And some of its munificence is meant for us. It is that prospect of what's to come that gives us the impetus to tend the ground, weed it of what is not useful, nurture the tender plants, and bring them to their fullness. The deep knowledge of what awaits keeps us vigilant and caring of our duties, nourishing the mystery of resurrection and renewal, greater by far—as in the Easter story of the empty tomb—than any stone blocking the way. Jump Time is a season of hope in which growth is called forth by the lure of the times of greening that are upon us.

But Jump Time has its weeds as well. We live in an age where every shadow is out in the open to be seen and felt, amplified by a media bent on entertaining us with horrors. The evening news can numb us into apathy or stir our spirit into action. How can we, in the face of negativity and collective fear, take the longest stride of soul to join the potentials of our local life to the Potency of the larger life that dwells within us all? And how can we do this so that what we have deemed extraordinary becomes the wonderful ordinary and the numinous extended universe takes up residence in our hearts and homes?

We must start by tending our own gardens. Personal strides of soul require time and space to generate the energy to move us beyond the shadows and losses of heart that keep us stuck. We need to give ourselves time to dream our future. We need the mindfulness to scan our day, our week, our year, and beyond. We need to pay attention to what we are doing when we feel ourselves to be most in the flow, when we feel happiest and most truly ourselves. It is here that the entelechy self can be felt and known. The entelechy self comes to us with the sense of what we would be like if we had spent a thousand years developing our full potential. I describe this feeling in my book *A Passion for the Possible*:

> Our Essential Self has a radiance that our local self does not. It is in touch with both our life and the Life of the Universe. It is in touch with the wisdom of the earth and the wisdom of the heart. It can put us in touch with the unexplored continents that lie within our minds and bodies, for it knows the maps of the soul and the treasures that can be found there.
>
> The Essential Self knows the possible paths our life may take and wants to help us choose the best ones. It knows how to turn imagination into reality and make the life we live fulfilling and creative. Above all, it knows why we are here and what we yet can do; where we can go and why we need to go there. (San Francisco: HarperSanFrancisco, 1997, pp. 92–3)

But to face the challenges of Jump Time, we must go further. As we come—and more and more of us are—to the recognition that our individual destiny and the world's unfolding are linked, we must also ask ourselves, "How best may I serve? How can I plant and nurture the seeds of a better world?" Then our strides of soul carry us beyond the landscape of our local consciousness into the fair country that opens into the entelechy of collective intelligence. There we are able to access the greater entelechy patterns that guide our world's social and cultural institutions toward their fullest expression in evolved social action, family and community life, governance, education, spirituality.

How this occurs is mysterious, but the study of creativity and inspiration furnishes many examples of people gaining access to ideas and blueprints that exceed their imaginings and study. One of my students, a talented musician, described such a moment this way:

> When I was in college, I simply knew that I was supposed to compose music. So I went about living a composer's life. Then, on the side almost, I started teaching movement and music to a group of young children, techniques that I had been using with actors and other professionals. That little class sparked a sequence of serendipitous and obviously planned connections—planned not by this local self but by something larger than me. Then a moment came, while I was teaching a parent/child class full of babies and three- and four-year-olds, that I just knew. My God, *this* is what I'm supposed to be doing! It was totally different from what I had expected my life to look like. So for a decade now, I've been teaching music to kids, and it just keeps flourishing. This work reaches me at a heart and soul level, beyond the inspiration, beyond the excitement, beyond the pleasure, beyond the raw energy of creating music and theater and all those things I used to think were important. There's a God-like thing to it, a universal thing. It's not what I sought out. It found me, and it sounded me. And once I hit that point, I've felt blessed.

And after all, why not? We are cosmic stuff, embedded in the ecology of the Whole, the universe in miniature, the unfolding of the divine seed, now in the springtime of its Self-consciousness. If anything, the muchness and interconnectedness of the modern world actually pushes us toward realization of our potential. Complexity and chaos can act as levers to lift the world that is too much upon us so we may see the patterns and information that lie within.

By harnessing our individual purpose to a vision of the possible collective future, we pull ourselves out of the mire of stasis and into tomorrow. Our spirits are called home, and we find heart for the next stage of our evolutionary journey.

CHAPTER 2

THE SELFING GAME IS WHAT INFINITY DOES FOR FUN

Recently, a friend I had not seen in a decade was to meet me at the clock in Grand Central Terminal. I wandered around for fifteen minutes looking for the familiar face and form of a woman in her fifties. Finally, a svelte and smartly dressed suburbanite in her thirties who had been smirking under the clock introduced herself as my old friend. Only when she spoke did I recognize the nasal voice and slight dentalization in her speech.

"Francine!" I exclaimed. "What has happened to you?"

"Well, Jean, I went all the way. My nose has been bobbed, my lips enlarged, my hips de-fatted, my hair replanted, and my face is a collagen factory!"

She said all this with a glow of satisfaction such as must have filled Tolstoy when he penned the last words to *War and Peace.* Sometime in the middle of lunch she got around to why, after all these years, she had wanted to see me.

"I figure now it's time to work on the inward me. And I hear you're pretty good at that."

When I gently suggested that perhaps a better plan might have been to start with the inward self, she remonstrated, "Oh, no, that never would have worked. After getting in touch with my inner child—or whatever I'm supposed to do these days—I would have looked in the mirror and seen this old hag and that would have been the end of it."

In her fantastical and expensive forays into body morphing, Francine is an extreme case. But she was also acting out of a tradition that serves a deep-seated need in human nature. We have always wanted to be seen as different from the way we are. I can imagine Francine, staring happily at her reflection: "Mirror, Mirror on the wall. Who does not look like me at all?"

If the mirror could reply, it might ask a question of its own: "Who is the who who's asking?" Who, indeed? Too many of us have been circling for years in the same limited notion of who we are and thus through the same patterns of neurosis. This is a sickbed psychology—one wound heals, and the problem pops up again in another form. It seems to me that Jump Time requires more—nothing less than a new self, one with a more flexible personality and the ability to laugh at the whole human comedy and at the self-identity we all take so seriously.

Since the Renaissance, Prometheus, the god who snatched fire from heaven, has been one of the Western world's reigning archetypes. But Proteus, the shape-shifter, is fast becoming the model for Jump Time. Whereas Prometheus stood for our seizing power over Nature and the elements, Proteus represents an even more drastic step—our gaining the ability to change our very natures and shift from self to self.

Proteus belongs to a long tradition that ascribes to the gods of many religions the power to change themselves into anything they choose: Zeus, in the course of his amorous adventures, became a swan, a bull, a ram, a serpent, a dove, an eagle, even a shower of gold. The Hindu god Vishnu is said to have been incarnated as a fish, a tortoise, a boar, a man-lion, a dwarf, and the Buddha. And, of course, Jesus was not recognized by his disciples when they met him on the road to Emmaus. Only over dinner was the spell broken. In the *Odyssey*, Homer tells how the soothsayer Proteus was seized, while sleeping, by Menelaus, who had come to him seeking knowledge of things to come. But Proteus was reluctant to give him any information and, to escape, turned himself into a bearded lion, a snake, a leopard, a bear, running water, and a tree. In spite of these metamorphoses, Menelaus wouldn't let go, until at last Proteus resumed his original form and agreed to answer his questions.

In a world moving through so many changes, the protean personality is the one that flourishes, for it has more points of view and thus more skill in anticipating outcomes. Francine's problem was that she did not go far enough. She had opted for the polish but not the pith of Proteus. Though her outward appearance had shifted, she still was the same old Francine inside, with the same old response patterns and emotional glitches. Yet the ability to shift identities, both outward and inward, is key to the *repatterning of human nature* that is giving birth to the new Jump Time self.

This new self shows up most obviously in the multitasking,

multimodal lives many of us lead. During a recent presidential election campaign, pollsters were kept busy charting the opinions of a new archetype, the "soccer Mom"—a busy professional woman running an office who also carpools her kids on their rounds of activities, participates in community life, and finds time to exercise, have a spiritual life, and spend quality time with her husband, if she still has a husband around. And if she doesn't, she's Dad, too.

In my conversations with these women and with many men who are also living "simultaneous lives," I find that most say they feel exhausted and fractured unless they move into another kind of awareness: seeing themselves as a theater of selves, a troupe of actors playing different roles according to the scene and script for the day. I have often suggested that if schizophrenia, the self that is split against itself, is the disease of the human condition, then polyphrenia, the orchestration of our many selves, may be our expanded health. This is more true than ever in Jump Time. The brittle raft of a solitary ego is an uncertain vessel in the sea of so much change. Polyphrenic people are able to keep a large cast of characters active, calling them to stage front as needed for the many roles they have to play. Each character brings new energies and a new set of skills. If people feel blocked or inadequate in some area, they move one persona over in the psyche, and then stand a good chance of finding an aspect of the self that is not blocked and is quite willing to tackle some dreaded activity.

In my local or habitual self, for example, I often feel downright stupid when it comes to writing. Despite this sense, I have produced a good many books as well as numerous lectures, articles, monographs, and seminar scripts. When a New Age type asks me, "Do you channel your books?" I am likely to retort emphatically, "I fight for every damn word!" Actually, what I do is to call on another persona of my inward crew who is not blocked when it comes to writing. I call this character the Working Muse and enlist her aid in the process of creation. A block of any kind is built up of habit, expectation, and self-fulfilling prophecy. It can become a massive chemical and protein complex in the brain and body. Instead of trying to defeat this formidable entity, one "switches channels" and brings in another character to whom the block is merely an interesting construct, a problem to be solved, not a well-worn track of sour memories and mental monoliths with the power to impede one's progress.

Where do these other characters come from? In what geo-psychic realm are they grown? Are they fantasy shapes or do they bear a reality that is thicker and more autonomous than imagination might have it?

From the depth probings of the psyche I have conducted over many years with people of different cultural backgrounds, I have arrived at a radical premise that is as mysterious as it is useful: I suggest that the many selves we contain are developed in three regions of the psyche. I call them the realm of "This Is Me," which pertains to the local and historical self in everyday reality; the realm of "We Are," the place of archetypal persona and the source level of principles and patterns of ideas and creative forms; and the realm of "I Am," Being Itself, or even God as the unity of all being. Doubtless there are many more realms, with levels upon levels within each, but these three stand out as significant, particularly as they relate to the production of our various personae. Like everything else in Jump Time, these realms of the psyche are undergoing their own systemic change and growth in form and function.

This Is Me

In previous writings I described This Is Me as the biographical region of the self, comprising our gender, family, history, local identity, profession, and relationships. I spoke of it as "the mask we wear, the persona of our everyday existence." Thus This Is Me is often hidden from its deeper identity in its essential or entelechy self and can exist between birth and death without ever finding out the greater purposes and resources it contained. But now in Jump Time, This Is Me drops the mask and drops the margins, becomes fluid, and flexes its polyphrenic muscles. Previously, the other selves that are part of our identity were mere glimpses, fleeting phantasms, more cinema than seen. Now, given the demands of Jump Time, they emerge from the mother matrix of the Self, a feisty family of selves with their own biographies, capacities, and requirements. This Is Me has become the consciousness of an abundant personal universe, with a formidable array of gifts and talents.

My work has brought me into contact with quite a few examples of successfully polyphrenic people, and what is so striking is the way their various selves feed and inform each other. Often this results in unusual abilities coming into play, including ones that exceed the norms of Western consciousness. I recently had occasion to telephone a young physician, Dr. Mona Lisa Schultz, who is a case in point. During our conversation I could hear the ringing of at least six different phone lines. "Those are my different professions," she explained. "Each line is for one of the things I am." The intermittent chimes and beeps that punctuated our chat attest to Mona Lisa's life as the ultimate hyphenated human: professor of psychiatry, neurobiologist, psychiatrist, lab worker, cat lover, author, medical intuitive, and intellectual wild card. The many hats she wears give Mona Lisa a genuinely holistic perspective, fey genius, and spectacular common sense.

"That's a Mona Lisa case," her colleagues exclaim when they run into walls over a client who stumps their expertise. Once, in fact, a client whose dilemma called for Mona Lisa's unique integrative perspective was actually a woman in the psychiatric ward who was always bumping into the wall. She had been an alcoholic and suffered memory loss. Her relatives thought her psychotic and hospitalized her. Mona Lisa studied her case scientifically, met her again on other levels of her personality, including the intuitive, and discovered that the woman suffered from Balints syndrome—a visual problem in which a person sees only a few things in the environment and not the environment as a whole. The woman could see only the tree and not the forest, thus her bumping syndrome.

For Mona Lisa this case is symbolic of so many of us who suffer from a lack of peripheral vision in our lives, marching with certitude into the walls of habituated thought and limited vision stemming from our blindness about the untapped potentials we contain. In the remarkable book that grew out of her experiences, *Awakening Intuition: Using Your Mind-Body Network for Insight and Healing* (New York: Harmony, 1998), she writes of intuition as a high art form and gives practical advice for orchestrating the myriad perspectives of our multicentric mind. Drawing on her own polyphrenic nature, she describes intuition as a natural function of our bodies; we are all wired for it, she says. But because we've shouted down its messages, intuition has to creep in the back door through dreams, physical sensations, and the unbidden aha's of recognition.

For me, Mona Lisa symbolizes the changing nature of personality, the coming Jump Time self, in which every sense organ is an agent for a different set of perceptions or ways of knowing. Mona Lisa, as may soon be the case for many of us, has simply tuned herself to alternative forms of receiving information and honed her ability to harvest this information from every part of her body/mind, from her personal history and, perhaps, from the collective mind as well. Like her, we are potent with the capacity for pattern recognition, a major factor in intuition. Then the loom of the mind activates and the shuttlecock of memory weaves old and new associations into a *Eureka!*

Another friend, Flemming A. Funch, founder of the New Civilization Network, described to me in a recent e-mail how life changes when we make intuitive polyphrenia a way of life. He writes:

> I've gotten used to operating in a space of uncertainty, steering by indicators provided synchronistically by the world around me. I used to have to plan things out and be very fearful of anything I didn't understand. But the world sort of overwhelmed me with its increased speed and unpredictability, and I had to adopt new ways. Now I somehow see trends as patterns amongst otherwise unrelated elements.

Among the many hats Flemming wears—computer programmer, futurist, Internet organizer—is what he calls a side profession as a counselor. Here, too, multimodal ways of knowing have replaced traditional skills. Flemming writes that he used to have a rigorous system of deciding what to do with people, planning out each counseling session in advance.

> But then I started finding that the people who walked in no longer fit my programs. I eventually threw them away, now

> just relying on a tool bag of principles and awarenesses. Somehow I magically always end up knowing what to do, and people walk away changed.

How, we might wonder, does the Jump Time polyphrenic self differ from the multitalented, multimodal person of the past? One might argue that running a farm in the nineteenth century involved as many different skills as those undertaken by the many-tasked person today. What with milking, baking, plowing, planting, harvesting, raising the children, raising the barn, vetting the animals, tending the poultry, mowing, haying, and caring for the sick and the needy, as well as being vigilant and resourceful in the cycling of the seasons—surely this life demanded multiple talents and varieties of the self. What differs in the polyphrenia of our own day is the disparity within the range of things we do. In life on the farm, all activities shared a common theme—Earth stewardship—allowing consciousness to maintain a certain accord. However rigorous the life, the mind was held within a necessary stability and even a sameness in which jumps were not encouraged. Then, too, religious practice held to its traditional containments, withholding ecstasies and other deviations. Supper conversation was predictable.

"How are the chickens, Clyde?"

"Laying good, Mama."

"Are you going to the church social tonight? Maisy Sue will be there."

"Then I'll be going."

Today, technology and the accelerated pace it engenders gives us a world in which we are absolved from the rule of seasons and the tyranny of cycles. My life and yours smack up against the unknown with a frequency that requires a mind that can morph with ease and multiply itself to encompass many frames and scenarios. Moreover, the varieties of ecstatic and far-reaching experiences now readily available have widened the spectrum of both choice and consciousness. Today's supper conversation proceeds along very different lines.

"Clyde, how was the Sun Dance with the Sioux?"

"Cool, Mama. I danced for four days and nights and had a revelation about the new physics as it relates to the vision of Black Elk. Did the video I made in Croatia about children of war come back from the studio yet?"

"Not yet. Tell me more about that Indian girl you fell in love with."

We Are

In Jump Time, it seems, our very nature is changing and growing in response to new complexities. The expanded self we need to see both trees and forest, to harvest the knowledge that is latent within to cope with the information that bombards us from without, cannot speak only one language. Intuition, multimodal ways of knowing, the complex counterpoint of the various voices of our polyphrenic natures are necessities today, if we are to survive and even thrive in Jump Time.

When brought into play these extended capacities are orchestrated by a Self that is liminal, that is, it can exist between the This Is Me world of space and time and the even more fluid realm of the We Are archetypes, where polyphrenia reigns. The mediator between the two realms is the entelechy, the guiding purposive self discussed in Chapter 1. The entelechy creates a bridge between the two realms, for it has access to the ways in which the local self is resonant with the universal selves and themes existing in the We Are. Say, for instance, a person who is involved in family counseling meets with a husband and wife who are totally at odds. Though the counselor listens to both sides and brings her This Is Me professional skill and experience to bear, the weight of the couple's emotions and projections blocks any progress. As the counselor prepares for her next session with the pair, she spends a few moments entering into resonance with her entelechy self—her absolute belief that her highest purpose for being is to mediate conflict. This moves her into the We Are realm where the energy of the Great Peacemaker can be found. She may associate this quality with the great peacemakers of the historical past or the archetypal ever-present: Gandhi; Francis of Assisi; Deganawidah of Iroquois tradition; Irene, the Greek archetype of peace. Gradually, she feels herself to be filled with the essence of peacemaking itself. When next she meets the couple, the counselor finds she is able to see beyond the immediate details of the conflict and weave the threads of reconciliation with a wisdom that she had not known before.

Since time immemorial the We Are realm has contained the energies of archetypal stories and myths, the tales of inspirational folk from history, as well as the great creative patterns that guide and instruct us. It also subsumes the panoply of numinous borderline persons, or "gods," who serve as rivers to the Source as well as vehicles through which we come to understand both our strengths and our shadows. Our new appreciation of the inherent polytheism of the psyche allows us to become more playful and more productive in the ways in which we approach what we used to think of as gods. In Jump Time, with new metaphors drawn from science and the spiritual technologies of various cultures, our relationship to the We Are realm is changing in ways that invite the wonderful as probable. We can learn to "download" programs from this region that were previously thought to be beyond our scope. Nobel physicist David Bohm has postulated a primary order of reality that is implicate, enfolded, and all-encompassing, just as the DNA in the nucleus of a cell encompasses the coding for potential life and directs the nature of its unfolding. In Bohm's terms, the implicate order is a grand frequency that contains in essence all other dimensions and knowledge. Buddhist and Hindu metaphysical systems also speak of a primary ground of being wherein all patterns and potentials exist. In contemplative or resonant states of consciousness, we can come into relationship with this primary order and receive the pattern, if not the details, of the knowledge we require.

Of course, indigenous peoples have always had access to such sources of greater knowing. It may be that the enormous interest in shamanism today serves as a subjective and vertical antidote to the limitations and arrested development of a life lived solely on the objective, horizontal plane. Furthering our fascination is the fact that shamanism, as the oldest form of religious life, is prepolitical. All religions began in spiritual experiences that eventually became politicized and bureaucratized. In both its ancient and modern forms, shamanic practice recalls a democratization of the spiritual experience in which hierarchies are predicated on levels of experience rather than on priests and popes. In shamanic practice revelation is direct, unmediated by the structures ordained by church or doctrine, and transpersonal dimensions of reality are available to anyone who makes the effort to master the techniques of the spiritual journey. Shamanic practice, however, has its dangers. Often it involves a radical disintegration and dissociation from both the ordinary life and the ordinary self, as well as a conscious entry into chaos. Living at their edges, outside or beyond themselves, shamans may experience ecstasy and rapture as a condition of mastery. At the same time, the ordeals of their voyages into shadow worlds sometimes harrow them in ways that few but trained shamans could endure.

Of course, like the rest of us, many contemporary shamans are living complex horizontal lives as well. I think in this context of Mamissi. She is also known as Mama Benz because she drives a Mercedes-Benz through the streets of Abidjan in the West African country of Ivory Coast. By day she careens through the city in her fine car, running many businesses, including food stalls, street vendors, and stores that sell the brilliantly colored cottons famous in that part of Africa. She is one of the biggest employers in the city and, many say, the most understanding and compassionate. But at night and all day Sunday she is transformed into a ceremonial priestess, a shaman healer dressed in flowing white. The sharp-eyed businesswoman becomes someone else. She enters trance states, communes with aspects of herself she believes to be "spirits," becomes deeply empathic with those who come for advice, and seems to know just what to do to effect changes in mind and body. Some of her techniques seem very odd to the Western eye. She chants and drums, cracks eggs on heads, spews gin on backs, and massages her clients with a mixture of crushed snails and aromatic oils. But mostly she serves as a bridge between ordinary reality and transpersonal realms.

Like shamans in all parts of the world, Mamissi has undergone an intense training in learning to journey inward to the place where psyche and cosmos gain access to each other. And what a journey it is! The shaman perceives a world of total aliveness, in all parts personal, in all parts sentient, in all parts capable of being known and used. This pananimism and panpsychism yields to the practicing shaman powers and principalities that can be used for healing, for renewal, and for bringing into the profane world the transformational powers of sacred space and time. Most important, like shamans the world over, she believes that the world is ultimately a unified fabric and that her job is sensing and moving with its ripples and waves wherever they lead—to communication with animals, spirits, and even rocks, until the healing path is found and the wisdom source is known. As anthropologist Joseph Epes Brown has said, speaking of the shamanically based Native American cultures:

> Unlike the conceptual categories of Western culture, American Indian traditions do not fragment experience into mutually exclusive dichotomies, but tend rather to stress modes of interrelatedness across categories of meaning, never losing sight of an ultimate wholeness." (*The Spiritual Legacy of the American Indian,* New York: Crossroads, 1982, p. 71)

The ability to self-orchestrate on the continuum of states of consciousness allows the shaman to gain information from parts of the self that most people have walled off from waking consciousness. A shaman may think of these different aspects as spirit guides, totems, and inner teachers, endowing them with power as they instruct her and fill her with their presence. As I watched Mamissi work, it seemed that she ran through a thrilling array of personalities that included the Great Snake, an African goddess, her grandmother, and the Embodiment of

Dance. Like Proteus, she is a supreme shape-shifter, in feeling if not in form.

From the distant past to the present Jump Time and across the boundaries of culture, the depth culture of the shaman persists, curiously consistent in both method and metaphor. Shamanic processes not unlike Mamissi's are used in a secular context by corporate overachievers to cope creatively with the demands of life on the edge. One well-known CEO moves through his day equally engaged in greeting touring school children in the morning as in discussing world issues with diplomats at night. In between, his tasks include lobbying recalcitrant congresspeople, talking to his dog, reading history, giving an emotional speech to a charitable foundation, playing a round of golf, analyzing budget figures, and figuring out ways to finesse his adversaries. Needless to say, his rivals call him the ultimate trickster and watch with grudging admiration as he juggles the many layers of his psyche, closing off some while he opens up others. Like a shaman, he keeps his balance by seeing the world as a unified fabric, each thread an integral part of an overall pattern of success and achievement. His mastery of his wide range of selves is vertical as well as horizontal, sourced in a belief system that imparts significance to every encounter and reads meaning across the categories of experience. The question of a child, a page read in the biography of a historical figure, or the remark of a dinner companion are equally likely to spark ideas for his many initiatives. Like most shamans, he has a shadow side, a potential for wildly inappropriate behavior that training and discipline could teach him to keep in check. In traditional indigenous societies, he would have received such training, but as a white Westerner he is without proper mentors. Nevertheless, he is prefigural, for psyche and cosmos meet in such protoshamanic people, and they point the way toward the heightened flexibility and deepened interiority that characterizes the emerging self of Jump Time.

I Am

The deepest realm of all, the one that engenders and energizes all the others, is the I Am. I see this realm as consonant with the assertion of God's Self-identity in the Old Testament, "I AM THAT I AM." In a literal translation of the Sanskrit text of the Bhagavad Gita, Krishna as Vishnu says words that express the character of the I Am as seen in the East:

> There is no past when I was not, nor you,
> nor these;
> And we shall none and never cease to live
> Throughout the long to-be.

The I Am is the sacred ground of being, the fount and origin of all and everything. Mystics experience it as a blazing union with Reality. In Jump Time, when we live daily life as spiritual exercise, the I Am become the uncommon commons, the field of Being wherein we dwell in the Kingdom and not in the outskirts. When we abide in the I Am, no longer do we persist in the pathos of the great divide between self and spirit. Our dependency ends, for we know the I Am to be immanent in ourselves, as we ourselves are partners in the enterprise of creation.

Myths and scriptures teach us that the gods are both One and many—a single unitary consciousness and its many refractions. Once in India I found myself in a village temple examining the rudely carved, brightly painted statues of the Hindu gods. A white-haired old farmer came over to me, his hands gnarled from years of plowing with his water buffalo.

"You see that goddess there, sister? That is Lakshmi. She brings us abundance and the fruits of the field. And over there with the elephant head, that is Ganesha, who brings us protection and many good things. Here in front of us is Shiva. He lets things die so that new birth can happen. And around us you see Parvati, Vishnu, Durga, Saraswati, Rama, and Sita. But really, sister, they are all *nama rupa,* that is, different 'names and forms' for the one God who contains them all."

Nama rupa is, in Hindu philosophy, the ultimate statement of the varieties of religious expression. The many names and forms of the gods give us some notion of how we might find unity and variety in ourselves. The gods manifest in multiple forms so they can best serve countless others according to their needs. But elevated into the One, they know Unity as their true nature. For us as well—multitasked, fragmented, and compartmentalized as we are by the shifting complexities of Jump Time—the one self that orchestrates the multitude that we contain is that part of the personality that is closest to the unitive nature, consonant with the divine.

What is more, the three realms of the psyche inform each other. When we embody the perspective of the I Am, the We Are realm of creativity and archetype becomes second nature, and we know the "gods" and the Great Stories to be parts of ourselves. This conjunction between the superpersonal We Are and the transpersonal I Am also affects daily life in the personal realm of This Is Me. For in the consonance of the three levels, we are primed by Being, empowered by story and archetype, and thus granted healing and inspiration in our personal life. To complete the circle, This Is Me, operating in the creative polyphrenia of Jump Time, becomes an ally and nurturer of the realm of We Are, which adds yet more novelty and discovery to I Am. In short, we have ceased to bore God.

The Selfing Game

The selfing game is what Infinity does for fun. And, given the profound changes in human psychology over the last millennia, the Infinite Personality has had a vastly entertaining time. For the most part, personality is patterned by one's particular historical period. In the past, people tended to develop their identities around tribes or communities, with individual differences and styles modulated to harmonize with the personality of the group. A weaver in fourteenth-century France belonged to a guild, lived on Weavers' Lane, knew his parents and remembered his grandparents as weavers, and was educating his children to be the same. Apart from a few drinks at the tavern and church on Sunday, his life was literally threaded on a single cultural loom.

Even when culture emigrated, it would, for a while, retain its shape. My Sicilian grandmother, Vita Todaro, brought Catania in eastern Sicily to "Brookalina," and there, on the stoop of her brownstone, she crocheted and sang the songs, served the food, spoke the curious dialect, and upheld the mores of a culture that had been her family's for a thousand years. It's a wonder that olive trees did not burst through the asphalt on Avenue O, though, to her credit, fig trees did grow in the backyard.

How different from Nana's sturdy sense of identity was that of my college classmates in the late 1950s! Here we were at Barnard from so many cultures—Jews, Italians, Irish, Poles, Russians, and a few WASPs—yet we had consciously stripped ourselves of the trappings of the Old World. Fueled by endless cups of coffee and by our passion for Jean-Paul Sartre, we anguished over the need to find an "authentic self." Jewish girls from the Bronx renamed themselves Solange, dressed in black, blew blue smoke of endless Gauloises out their noses, and actively cultivated an appearance of ennui.

Today, we go even further. Remaking ourselves, even many times, seems perfectly natural. Our "second" natures have third, fourth, fifth, and many more natures waiting in the wings to enter the front lines of our personality. Newpapers are filled with evidence of these multiple incarnations: a former Symbionese Liberation Army bank-robbing radical is discovered to be hiding out as an upper-middle-class doctor's wife in a midwestern suburb. An affluent real estate lawyer abandons job and family to shave his head and become a wandering forest monk in Thailand. A professional wrestler trades his feather boa for horn-rimmed glasses and the governor's chair. A polite Mormon teenager builds a billion-dollar Internet company and then, after an extraterrestrial visitation, gives it up to devote his time to advancing technologies known primarily to readers of science fiction. Women run corporations while men tend children, cook meals, and keep house for their high-powered wives. The news media furthers this process of self-creation by the images it beams around the globe. At a school in Bangladesh, a boy told me he was going to be an astronaut, while a girl assured me of her plans to invent ways to control the weather so that her country would not be ravaged by yearly floods. Just a decade ago neither of these children could have conceived of such ambitions, much less had any possibility of achieving them. CNN, it seems, has amplified the agenda of the self. This is not simply a case, as in times past, of children striving to go further than their parents. Rather, it seems, the self today is in search of its wider identities. Perhaps, as philosophers and metaphysicians have long argued, there is no "authentic self." The self may be a process, arising anew in each moment as a result of what we see, experience, discover. Perhaps the selfing game we are playing now with such gusto will bring home to us what the sages of the East have long held—that the "I" we cling to so adamantly and protect so ardently is a fiction; that in an absolute sense, there's no "one" home!

Interestingly, the new technologies of Jump Time seem to reinforce the idea that the self is what we make it in the moment. The Internet, for instance, allows us to explore a variety of latent identities and exist everywhere and nowhere in a realm where nothing is forgotten yet everything changes. Cyberspace is a parallel universe not bound by the usual categories of space and time. As a researcher into human capacities, I see the Internet as a phenomenal jump in the selfing game. Its Earth-wide neural system has the potential to create for us, quite literally, a new brain and a new body. Because it grants us nearly mythic experiences, the Internet activates the growth of extraordinary new capacities. We humans have always been prosthetic beings, extending our arms and legs through wheels, tools, machines, and our nervous system through electronic devices. The collective unconscious coded in the electrons of cyberspace is fast becoming the prosthetic extension of the human psyche and the catalyst for unparalleled growth in our interior database. Before now, few have accessed this inner realm of pure potential, partly because the metaphors were not present to release the code and turn on the system. I propose that the Internet provides that metaphor. The fact that the Internet works so outrageously well and in so fascinating a manner provokes our interior intranets to grow in parallel ways, mirroring the ever-increasing capacities and complexity of the Internet itself.

What happens within the psyche when it is hooked into the World Wide Web? Among other things, alter egos abound—the polyphrenic self interacting in a bouquet of identities with people who bring their own garden of selves into the conversation. Here, as elsewhere, science fiction anticipates science fact, and even the marvels of comic strip heroes prefigure life on the Net. Clark Kent is no longer confined to the phone booth for his transformation into Superman. The computer is a smaller and better box within which we can radically change identities and take on curious strengths. Faster than a speeding ion. More powerful than a mainframe computer. Able to leap continents with the tap of a key. Is it a bird? Is it a plane? No, it's Cyberself—the current alias of many of us who have morphed with our modems.

I know any number of relatively quiet middle-class men and women who live predictable and even humdrum lives who, after dark when the kids are in bed, join the Net as Champagne Baby, Thunder Boots, Magic Mountain Man, or Mega Mama. They scheme, they dream, and sometimes they even do. Freed from their local selves, the Net and their new identity empowers their passion and accesses their creativity. A fourteen-year-old girl in New England with a passionate desire to clean up the oceans has assumed an oceanic name, Neptuna. She has mobilized hundreds of people on the Net with similar concerns to band together to assist in this enormous task. Experts have been drawn to her, as have people with the will and the wherewithal to act. Most think of her as a mature woman and would be stunned to see that it is in fact a teenager tapping away on a home computer who is giving them suggestions and inspiration.

Where is this all going? Sherry Turkle has written, in *Life on the Screen* (Simon & Schuster, New York, 1995), about the ways in which the Internet is evolving a new sense of identity—multiple, virtual, Shakespearean in its ability to deconstruct personality, creating gender double agents—men playing women who pretend to be men, and women playing men who pretend to be women. In the chat rooms and MUDs (Multi-User Dungeons) of cyberspace, we push the envelope on the ways gender is constructed and expand our emotional range until the old conventions for interaction between men and women are reformatted into entirely new programs.

Shadows loom large, of course. The embodiment and broadcasting of private images across the Internet has allowed monsters to get out there, private deviance to have an audience, obscenity to flourish. The old-fashioned vices, electronically amplified, have never had it so good. Porn is ever-present, e-gambling holds millions in thrall, and many ordinary folk lose their savings to the addictions of electronic commerce, cybermalls, and auctions for things they never knew they wanted until they popped up like sirens on the Net. Recently, someone offered a kidney on the eBay auction, and only when the bid reached six million dollars did the site's managers realize what had happened and withdraw the "item" from sale. Should he want to buy a few million souls for his realm, the devil himself could take lessons from the Net marketers.

Despite its shadows, the Internet nevertheless is our most promising road to transcendence. We are today in the early stage of being able to assume avatars. In Hindu thought, an avatar is a god or god force who has downloaded into a human body and consciousness in space and time. Thus, Prince Rama in the great Indian epic the *Ramayana* is the avatar of the creator god Vishnu, and his wife, Princess Sita, is the avatar of Vishnu's wife, Lakshmi, the goddess of abundance. In cyberspace, an avatar allows us to take on and be seen on the screen wearing a virtual body: Merlin or Wonder Woman, Queen Elizabeth or Joan of Arc, a creature with a thousand eyes or a fish with the head of a monkey. Such virtual identities are closer to impersonation than incarnation, yet some people come to identify so closely with the avatar they wear that it becomes an extension of their human personality, allowing them to express feelings and points of view in poignant ways. A scientist who in actual life comes across as a tough cookie in the laboratory in the avatar world might wear an Earth head to express the powerful love she feels for all Gaia's creatures. The avatars of cyberspace allow us to try on the diversity we contain and to interact with other avatars in virtual landscapes that bring the dreamworld onto the screen as a catalyst for the psychology of polyphrenia. In Jump Time, with the help of the Internet, our inner palate will widen even as our outer experiences reflect the greater range of cultures and peoples we encounter at work, in our communities, and in our own families.

In times past, people would wear the masks of the gods on ritual occasions. Then they would act as if they had a god's abilities and powers. Cyberspace has given us the opportunity to imitate our ancient brothers and sisters. We are learning to take on archetypal identities and experience the world in more powerful ways. We are growing a more fluid personal psychology and, with the input of the Internet, changing the way we think of ourselves. Maybe in Jump Time, interconnected and intraconnected as we are coming to be, the notion of the self as we currently conceive it will disappear.

The World Self

Let's try a little time warping. Imagine that you are a student of planetary histories several thousand years into the future. You have taken a sampling of a hundred planets on which sentient life has achieved a high level of consciousness. You observe a characteristic pattern of development, a variety of evolutionary progressions that all result in a dominant self-conscious species. You note that only when certain thresholds are reached does evolution accelerate. For example, when the population of the dominant species has reached a critical number, the male and female achieve equal status and the rich mindstyle of the feminine transforms all institutions, all sensibilities. Added to this, when technology reaches a state that allows for intensive interconnection, the species makes strides toward ending its fierce and ancient enmities.

Then something remarkable happens. On many of these planets, this interconnectedness extends to the planet itself, and feeling arises in the species for the entire planetary organism, a caring and mutuality in which, it seems, the living planet embraces them as well. A new order of friendship is born, one perhaps known in the distant past by those who revered the planet and welcomed her largesse in rite and celebration. But now this emotional bond is internalized, the law written on the heart, the planet's soul consonant with its species, the recognition of a seamless web of kinship between all planetary creatures, the return of deep ecology.

Now things really speed up. Latencies that had been known to a few become the province of the many. The organs of sense refine and become more inclusive. The modalities of knowing expand. The species is earthed and skyed. Ancient entities once thought to be the embodiments of planetary forces move inward. What were once "gods" are known to be innate, and inwardness becomes as vast as outwardness. The insular self metamorphoses into a continuum with the world and develops planetary consciousness. Then history as it had been known comes to an end, and another kind of story ensues—the story of cocreation.

Like all healing fictions, this story may not be accurate, but it is true. For it seems clear that there is a pattern, a plan even, enshrined in our cosmic coding, our psychic DNA, that comes alive in us after certain requirements are met. Then our dormant "world organs" awaken. The personal ego, recognized to be a cultural and political artifact, becomes transparent. The change is textural rather than structural: where ego was, entelechy comes to be. Entelechy always encourages polyphrenia, more selves to carry the abundance that comes with extended senses and knowledge. Entelechy leads us inevitably to the personality of the future—what I call the world self.

The world self has another story—the unfolding of the cosmos. It recognizes that personal history is a subset of universal history, and local time and space are blown open to a more expansive domain. Entelechy operates to give us access to the deep architecture behind things, the realm of great pattern and creative form that was and is and shall be. Tasting these riches, we return, sourced with ideas and with the passion to reform and reweave our community, our world, and, most important, ourselves.

Mystics and high creatives know this secret. Each enters a realm where they encounter with crystalline brilliance the very forms of reality. The difference between them is that while mystical experience is union and immersion *in* the Source, creative experience is gaining information and generative potencies *from* the Source. In both, there is awakened sentiment of love, exhilaration, and awe. The barriers of local personality are breached. One builds bridges to the cosmos—then looks back on the self from there. Life is enhanced, the soul of the world is known, and one ascends to the apex of one's spirit and what has been called the extended life of the All.

Such cosmic consciousness is innate to the human condition. When we see the world as a living being, an extension of our own consciousness, we stop projecting onto it and labeling it accordingly. And it's about time, too. The world as projected by our limited conceptions and our fears has led to one holocaust after another, the psychosis between cultures. Vital, buzzing, flaming sensory richness, both real and imagined, incinerates the flimsy tissue of concept. Labeling crumples to ash in the fires of immediate perception.

A bird has just flown into my windowpane. It has, I believe, knocked itself out. I can dismiss the whole event and go back to my book, or I can remember the sound of the bird hitting the glass and the fall to the ground, imagine its wing broken, and wonder if it was gathering worms to feed spring chicks warming under its mate's feathers. Spurred by these sensory imaginings, I go outside, find the bird, chase away the stalking yellow cat, and discover that it is stunned but unbroken. When I warm it in my hands, its little heart flutters, and off it flies. Concept alone would have blocked my emotional response and left the world one bird less. Take this little story to another plane, and you have the trauma of the twentieth century—our tendency to see people as ciphers and statistics to be manipulated, the feeling tones of their lives unremarked and disrespected.

The world self is fed by direct perception, by the dissolving of barriers between self and other. When our local self comes into resonance with what Native Americans call "all my relations" and Buddhists call "all sentient beings," we act in ways that are consonant with both our highest purpose and the world's destiny. This enlarged perception brings us into phase with the evolutionary wave that is propelling us and the universe forward. We escape the sense that our choices are limited or that life is without meaning or purpose, and our sense of the future opens.

A close friend wrote to me of her own experience of such a moment.

> One early morning, after meditating and before work, I decided to take a walk around a nearby pond instead of doing my usual practices. I'd been feeling out of balance and thought the air might clarify things. As I walked around the far end of the pond I suddenly (but not abruptly) saw several streams of light extending out in front of me. In each light beam I saw, with perfect clarity and definition, myself living out a life in the future.
>
> In one I was married and living in a house with a fireplace, in a sweet and serene domesticity. In another I had given up my regular job and was doing freelance choreography for theater and dance productions. In a third I had decided to devote myself to the college dance program I direct at this time, devel-

> oping and strengthening it within the college. And in yet another I had left the East Coast altogether and was living as a contemplative.
>
> There were more, which I didn't have time to peruse, because as the geese stopped honking and the mallards stopped quacking, an enormous great blue heron flew easterly down the pond and lit just several yards ahead of me. I had not before (nor have I since) seen a great blue on that pond. I felt tremendous support from the natural world, from the visionary world, and from my inner life all at one time. My sense was that any of these parallel lives would be fine, would be perfect, should it wind up being my whole life, the rest of my life, my only life.
>
> I have since that day, more often than not, been able to see each moment as simply what I'm doing. Not a drama, not a defining construct, but just what it is. What I'm doing at that moment becomes everything, and all time. There's a sense of both sufficiency and possibility.

As my friend discovered, when we embrace the richer perspective of the world self, the psyche is remade. We experience a more natural joy, an acuity of feeling for the Earth and for others; we enter into the space of Unity and experience the divine play of being and becoming. It is like becoming a million years.

CHAPTER 3

WITH SATCHEL AND SHINING MORNING FACE

Consider the problems today's young people will have to face: global warming and other changes in climate, disastrous ups and downs in our interlinked financial markets, worldwide unemployment, more than a billion people living in deprivation, disappearing soils and forests, oppressive governments and corporations, a stratified economic system that rewards the most greedy among us. As we confront the challenge of *repatterning human nature* to succeed in Jump Time, we must ask ourselves: What kind of education do we need to develop the skills to cope with a world in which so much can go wrong?

It would be the greatest of tragedies if now, in the midst of the golden age of brain/mind research, when we are discovering the full range of what we contain and what we yet may be, we agree to a limited vision of our possibilities amped up with new technologies. Education that is hands-on, sensory rich, and experience laden, that calls forth the whole mind of the whole child, can develop our human potential and give us the tools to cope with whole system transition. Moreover, schools can lead the way in providing the model for education that is continuous throughout life, so that adults can be and know and discover human wonders that exceed even the most far-out technologies. This is what schools at all levels can teach. In so doing, they prepare us to be possible humans who are equal to the task of navigating the shoals of Jump Time.

Perhaps by looking at an earlier Jump Time and its optimal education, we can discover ways of reinventing our own.

Shakespeare at School

Why is it, I wonder, that as the millennium begins, a film about Shakespeare entrances and engages us, and Shakespeare's plays, written four hundred years ago, are the backbone of local theater groups as well as movies, novels, and Broadway productions? There is mystery here, and magic as well. Perhaps across the centuries, something is calling us to look to the past to discover our future.

Just saying his name, "William Shakespeare," sets us off. He is a whole mythology unto himself and a continuous industry for scholars, teachers, and filmmakers. He looms out of the Jump Time of Elizabethan England with an astonishing abruptness—

genius, of course, is always abrupt and unexpected. But this is a man from someplace else. For how on earth and under heaven could such a mind of incomparably greater powers than anything that had gone before (with the possible exception of Leonardo da Vinci) have so mysteriously appeared, if not from someplace else or some great unbounded cavern of the soul that few have ever visited?

Perhaps some answers can be found in his education. Indeed, his education may give us the future of our own, for in its own way, it was prescient of the best things we know about drawing forth the passion of our minds and the fullness of our spirits. So let us discover Shakespeare as a child. There is Will, up at five, wandering to school through dawn and bird song, "the whining schoolboy, with satchel and shining morning face, creeping like snail, unwillingly to school" (*As You Like It* act 2, scene 7, lines 145–147). After he has learned his ABCs out of the hornbooks of the time, it's on to grammar, to memorizing reams of poetry, hearty maxims, whole books of the Old and New Testament, and the ever-present Latin—Cicero, Cato, Aesop, Horace, Virgil, and the glorious ancient Roman playwrights Terence and Plautus.

And he acted the plays as well. Richard Jenkins, who was most likely his major teacher, had been a student of Richard Mulcaster, the headmaster of the famous Merchant Taylors School of London. Mulcaster believed that children could learn best if they acted, and he regularly brought his boys to court to act in plays. Acting, he insisted, would have a good effect on their "bearing and audacity," for he knew the uniqueness of the time in which he lived: "This period in our time seemeth to me the perfectest period in our English tongue. . . . There is in our tongue great and sufficient stuff for art."

So little Will was a performer from the start. Theater was the way his mind was formed. In addition to being in plays, he frequently saw them performed; Stratford, as a market town, was visited by most traveling players and mummers. Thus from his earliest impressions Will discovered that "all the world's a stage."

Then, too, the world that he lived in, sixteenth- and seventeenth-century England, was the most literate society the world had ever known. Fifty percent of men in cities and 40 percent in the countryside could read, and women were not that far behind. England seethed with new ideas, and the public participated in debate and learning on every front. The monarchs, particularly Elizabeth and James I, were more involved with scientific and literary as well as classical explorations than any rulers before or since.

Thus it is no surprise that in this period of unprecedented intellectual vitality—a Jump Time not unlike our own—the demanding plays of Shakespeare were popular with a large and wide-ranging audience. His words indeed "split the ears of groundlings," who would applaud and shout with delight. Shakespeare's words and uncommon wisdom were birthed in that school in Stratford, a place that was a product of educational reforms suggested by the quintessential German educator and humanist of the early sixteenth century, Desiderius Erasmus.

Humanism was an attitude of the mind that accompanied the flowering of the Renaissance—the *studia humanitas,* those studies of grammar, rhetoric, poetry, history, and moral philosophy thought to possess the ability to make a fully realized human being, a new order of the possible human. Erasmus believed that the mind could be engaged in fresh ways though translations from one language into another. And so a grammar school boy of Shakespeare's time spent eleven hours, six days a week, for up to ten years, translating from Latin to English and back again, analyzing Latin literature, much of which was based on Greek stories and sources, reading aloud, reciting, and memorizing.

It's no wonder, then, that when Shakespeare got to upper school, he fell in love with the great Roman poet of myth and symbol, Ovid. As much as the Bible and the Book of Common Prayer, Ovid fertilized his mind, touching off deep memories of mythic power that gave him some of his finest plots, subjects, characters, and themes.

Elizabethan England was also the all-time place for memorization, often through hearing things spoken, as books were hard to come by. Shakespeare's aural memory was prodigious. His vocabulary was saturated in the grandeur of words and phrases that one finds in Latin. With so many thousands of poems and word combinations in Latin, Greek, and English roaring through his mind, he borrowed from everybody, and his great speeches are a remade world of submerged and unconscious literary memories and quotations. Shakespearean scholars have waxed fat and foolish trying to trace his indebtedness to sources, but they lose sight of the fact that Elizabethan culture was a great stewpot of language, memory, and literary pyrotechnics. Just as Renaissance Italy was the premier center for visual images, so England became supremely the place for auditory images. It is the nature of images to spawn other images, and auditory images weave and wind sounds together to make for new conjunctions, indeed, for new thought.

Erasmus further counseled that students learn the art of *controversiae.* He advised that students engage in debates, devising arguments, both rational and emotional, to persuade their hearers that first one side and then the other is correct. Debating taught students to argue both sides of a question convincingly. Perhaps this is why Shakespeare's own opinions are never completely clear. He is "myriad-minded" and can so successfully persuade us that several views are equally correct that we cannot find within them any that can be said to speak with his authentic voice. In this Shakespeare is the genius of polyphrenia. He creates his characters as human beings and sets them free of obvious manipulation.

Another formative practice in the training of young scholars in Shakespeare's time was imitation, a writing exercise that is neither slavish reproduction nor mere translation. One studies a great piece of writing by one of the acknowledged giants of the past, enters into a process of internalization—an alchemizing through one's own life and experience—and then creates a poem or other work that is unique to the writer yet has similarities to the original. This practice enriches one's ways of thinking, deepens one's ability to allude to other forms, thickens the soup of one's mind, and begins to make it possible to play, as Shakespeare and his contemporaries did, a delicious and pleasurable game between literature and life.

Perhaps the most important of the educational demands of Erasmus, and widely used in Shakespeare's day not only by schoolboys and scholars but by gentlemen and courtiers, was the notebook. Divided into sections, with themes and categories, one noted in it everything one read, saw, and heard. We know that Leonardo da Vinci kept copious notebooks, adding visual notations to his ideas and observations. It's as if he were taking notes for God. Closer to our own day, anthropologist Margaret Mead always carried a thick, square red notebook in which she jotted each day things that interested her. She generally filled one of these books each month. Interesting ideas, evocative turns of phrase, emotion-laden scenes, research materials, and her considered reflections on these all ended up in her red notebook. She told me that every night or early in the morning she reviewed the day's observations to keep them fresh and fertile in her mind. The Library of Congress now has many of these red notebooks on file. There is a vast difference between this kind of precise written observation and what we call journaling.

With so much to draw on in his inner library as well as his notebooks, Shakespeare experienced an astonishing ease and rapidity of interchange between literary texts and the life of spontaneous feeling. He was a world-maker, creating a world, populating it, and setting it to spin. He kept his characters in his mind all the time and was aware of what had happened to them before the play began and what was happening to them at each moment within the play, even when they were not on stage. Thus they continued to develop, even when he was not holding the quill. Perhaps that is why Shakespeare's creations never really die, why Romeo and Juliet, Hamlet and Lear, in spite of their tragic endings, continue to live in the mindfield, where Shakespeare planted them for immortal life.

The Clara Barton School

William Shakespeare is a Jump Time in his own right. He is the soul of his age, the supreme educator, who blends his conscious art with a cornucopia of sources so that his audience and readers respond with all that they are and more than they know. Great teachers do this, and great education—for all its differences in content from that of Shakespeare's day—brings the mind of every child into congruence with its own genius. I have observed many schools and many styles of learning the world over, and the best of them, the ones to which children run in delight and expectation, are those where learning is creation, performing, thinking across subjects, exploring ideas through images, sounds, songs, dances, and artistic expression. There children become, like Shakespeare, "myriad-minded"—conscious participants in their own unfolding. Yes, they continue to read, and write, and cipher, but they are also encouraged to imagine, dream, and expand the limits of the possible.

A strong connection links little Will in sixteenth-century Stratford and Willy in twenty-first-century Minnesota. At the exceptional Clara Barton School, the finest characteristics of Shakespeare's education are realized in brilliant contemporary forms, with wonderful results for children, teachers, parents, and community. Looking in depth at this model school gives us a good picture of the best the current educational system has to offer.

Willy does not translate from English to Latin and back, but he *is* transferring what he is learning in math to what he is studying in music—the progression of harmonies made clear in their mathematical arrangement. As we now know, music lessons enhance spatial intelligence—crucial for engineering, computational abilities, and technical design. Willy corresponds via e-mail with Vikram in Madras, who is also showing considerable ability in both math and music, as are many of his classmates, for the ragas that fill the air in India have complex rhythmical and tonal patterns that call forth the geometries of the mind, the algebras of consciousness. Willy's school emphasizes music and art, because its teachers know that arts kindle the imagination, stimulate the brain's connectivity, and give students firsthand experience at world-making. As Shakespeare knew, and as the best schools of Jump Time put into practice, the arts make us human.

The more we learn about the ways students learn, the more important arts education becomes. Recent research indicates that less than 15 percent of students are auditory learners, that is, they process information primarily through hearing it. Visual learners, who process information primarily by seeing pictures, account for 40 percent of students. Kinesthetic learners, who respond best to hands-on learning, are the largest group; fully 45 percent of students need immediate sensory stimulation to learn effectively. Kinesthetic learners often have trouble in traditional verbal-based classrooms, where opportunities for high-touch learning are rare. Arts help visual and kinesthetic learners—in fact, all students—to learn more quickly, retain what they have learned, and feel more positive about learning.

Recent research shows that if children sing songs, they learn math and languages better. The mental mechanisms that process music are deeply entwined with brain functions such as spatial relations, memory, and language. Give children singing lessons and keyboard instruction, and their mathematical abilities soar. At the Northwest School in Seattle, students in grades six through twelve are required to take two arts courses at all times. Offerings such as dance, drama, music, and the visual arts are taught by practicing professional artists who are also strong teachers. Frequent informal performances and exhibits encourage students to feel proud and confident about their work. When nine college admissions directors and independent college counselors were asked to rate local high schools, the Northwest School was ranked second overall among Washington high schools and highest among private high schools in student achievement during their graduates' first year at the University of Washington.

Music instruction even helps students learn science, as is demonstrated by high science achievement scores for eighth and ninth graders in Hungary, which up until recently had the most intensive school music program in the world. Dance energizes and stimulates the entire mind-body system. Another study of 250 elementary students showed that they improved significantly in language arts when movement and dance activities were expanded, with test scores rising in correlation to the amount of time spent in movement activities.

Results like these underscore why arts programs are so critical. A child can learn math as a rhythmic dance, and learn it well, for rhythm is processed in the brain in areas adjacent to the centers for pattern and order. A child can learn almost anything if she is dancing, tasting, touching, hearing, seeing, and feeling information. She becomes a passionate learner who delights in using much more of her mind/brain/body system than conventional schooling generally permits. Much of the failure in schools stems from boredom, which arises from the system's larger failure to stimulate and not repress those wonder areas in a child's brain that give her so many ways of responding to the world.

At the Clara Barton School, Willy's classroom buzzes with rich language. The combustible words of great literature come pouring out of the mouths of students and teachers alike in the "read alouds." Students are inspired to try their hand at writing in the style of the literary greats, identifying the techniques and using them in their own poems, plays, and stories. In third and fourth grade classes, students read and compare versions of the

Cinderella story from many cultures, identifying common themes, exploring cultural differences, and then writing their own versions. Willy's classmate Marvella discovered that, like Cinderella, Oprah Winfrey was mistreated as a girl by members of her family. Then she looked at the life of Nelson Mandela and saw in his childhood an African Cinderella story. Finally, she wrote her own version, set in an urban ghetto. Mirroring techniques are also used in the visual arts. Children are encouraged to use the line drawings of Picasso, Dürer, and Degas as stylistic models for their own sketches. Imitation of art enhances fine hand-to-eye coordination and gives children extended powers of observation, a skill needed in many sciences.

Currents of theater and drama flow through all levels of the curriculum at Clara Barton. Children in the primary grades act out stories and poems in order to interpret them and practice social behaviors by dramatizing a full range of social conduct, from the comically disastrous to the elegant and courtly. By the fourth grade, theater games are a constant, and history class is frequently enriched with reenactments of the signing of the Declaration of Independence or the delivery in character of great speeches of the past. Students of all ages take part in theatrical events, putting on everything from *Guys and Dolls* to Shakespeare to their own plays, including an original musical, "Science, So What?" The school is also a veritable Stratford in that traveling troops of players regularly perform for the students. In a recent month, students were treated to a one-person show about Jackie Robinson, a new play weaving a Japanese tale with issues of Asian adoption in Minnesota, and the famous puppet theater Bread and Roses. The school's annual fundraiser is also a full-dress theatrical occasion at the local weekend Renaissance Fair. Parents as well as students are caught up in the spirit of the re-creation, and many come to school in full Renaissance dress, acting the parts of artisans, merchants, and noblemen and noblewomen, telling stories, teaching crafts, and enacting thrilling duels.

I am a passionate advocate for the use of theater in the classroom. In theater, the child becomes the possible human, using all skills—music, dance, rhetoric, expression, feeling—to tour the landscape of human experience. If all the world's a stage, then all stages of life, all grades of human aspiration, all levels and layers of human expression and emotion are scaled when drama comes to school. Walt Whitman once said, "I become what I behold." Childhood is that special time when the margins of the self are leaky. Theater allows children to try on the many parts of the human comedy and the full range of human knowing. What is more, what one enacts, one remembers.

Debate is also high on the academic agenda at Clara Barton. Children in the middle grades practice a contemporary form of *controversiae* and are provided with many opportunities to see all sides of an issue, speak with different voices, and express opposing views. Recent debate topics in Willy's science class span the spectrum of challenging and charged contemporary issues: the ethics of cloning, surrogate parents, the use of fetal tissue in research, organ transplants. Recently, Willy's history class debated whether the U.S. government or an international UN force should try to capture and kill Saddam Hussein. In the midst of an emotionally laden debate, Willy's teacher often asks students to switch sides and argue the opposing point of view.

Words on the page are stimulated by the brain in motion. Metaphors are the stuff of the brain's transformings. Take time to observe an approaching fog, and it becomes, as in Carl

Sandburg's vision, a creeping cat. Dancing to Saint-Saëns' *Carnival of the Animals* translates into rollicking images of how it feels to be large and lumbering, furred or fanged. The rough texture of tree bark branches in the mind to poems that root in earth and reach to touch the sky. But even more, the mind grows a network of images and ideas and charges these with the chemistry of creation, and what was seen, heard, imagined is then planted on pages redolent with discovery. The child writes and rarely fears to write again.

Starting very early, children at Clara Barton learn to savor personal writing. Like Leonardo's, their journal notebooks have sections for writing, drawing, and deeper reflections. Life becomes an opportunity for making a record of day-to-day happenings. A second-grader came bursting into her classroom one morning in great excitement. "My friend broke her arm, and I have to write about it in my journal," she exclaimed. In Launa Ellison's fourth-grade classroom, Kyle creates in his journal a continuing cartoon strip with original characters, Kaelyn writes a unfolding story with weekly episodes, while Steffanie is pouring out poetry.

Though the Clara Barton School takes part in district and state required tests, few teachers use exams as part of the classroom learning process. Instead, students in every grade keep two portfolios, a yearly collection of work and a pass-along collection that follows them from grade to grade. Younger students like Willy come with their parents to lively Portfolio Parties, where students are encouraged to present their work to their parents. In preparation for the event, students choose a sample from the art, math, science, and writing sections of their portfolios for posting on a "best work" bulletin board. By the seventh grade, students share their portfolios in high energy performance evenings, filled with joy and laughter, discussion and creativity, that allow students both to demonstrate and to reflect on how much they have learned. Each May, children and teachers from kindergarten to the upper grades choose three to five works from the year's collection to place in the students' pass-along portfolio, which provides continuing evidence of the changes they have made. When Willy graduates from the eighth grade, his parents will be presented with this cumulative portfolio as part of the closing ritual of his years at Clara Barton.

New Skills for a New World

The kinds of learning experiences offered at the Clara Barton school and institutions like it would be wonderful if widely applied, not just in North America but in schools all over the world. They represent the best of what we know about education for the twentieth century. Unfortunately, much of our educational system falls woefully short of this high standard. All children need windows of opportunity that will open them to a world that they will partner in new creation. Never has the time been so ripe for change, and never before has education been more available to new strategies. So much is now known about the nature of learning that, if applied broadly, could potentially transform civilization and create, within several generations, people who are endowed with the skills and moral courage to navigate in Jump Time.

Neuroanatomist Marian Diamond at the University of California at Berkeley has shown that the human brain can change structurally and functionally as a result of learning and experience—for better or for worse. When we are in environments that are positive, stimulating, and encouraging of action and interaction, the human brain continues to develop throughout our lifetime, growing new neural connections that enhance our capacities for learning, creativity, and problem-solving. In effect, we make our brain as we use it.

In the Jump Time of the new millennium, we must use what we know to educate ourselves for the next civilization, the one that exceeds our expectations. In a sense, what is needed is training for the unknown and inexplicable. We must discover ways to "cook on more burners" and to democratize skills that in the past belonged to the few. This challenge takes schools into areas of human potential that assume as a given the education of the whole mind and body that occurs in forward-looking schools like Clara Barton. It also spurs schools to embrace capacities and sensibilities that traditionally belong to mystics and mages, high creative folk and world servers.

The goal of this kind of education is what I have called the possible human, the fetus of the emerging self. Perhaps this full expression of human potential has been coded in us since time out of mind, but only now, given present complexity and crisis, have so many been called to its realization. What skills this possible human can evince, what use of the tremendous palette of our given but forgotten nature it implies, may be key to our stewardship of the Earth and to our continuity as a species, as well as to our future as star-goers, explorers, and creators in a universe both real and visionary. Let us look at some of the skills we might need to grow the possible human.

Body Skills. The first group of skills is engendered through an emphasis on dance, sports, and physical training, giving children an enjoyable experience of being in their bodies, flexible joints and muscles, and a physicality that is fluid and full of grace. A fine tuning of body senses and physical abilities assures a greater appetite for delight and for the pleasures of being human—the sheer enjoyment of being corporeal.

The stimulation of a multisensory education gives children acute senses, which are not limited to five, for they enjoy and maintain a state natural to most children: synesthesia or cross-sensing, the capacity to hear color and touch the textures of music, capture with their noses the smell of words, and taste the subtlest of feelings. Since their sensory palette is so colorful and wide ranging, if encouraged, children can begin the journey toward engaging the world as artist and mystic, seeing infinity in a grain of sand and heaven in a wildflower. This perceptual style can blossom into a capacity to recognize the patterns that connect the forms of life and thought to each other. It grants an acuity of observation that gives meaning to randomness and the ability to see emergent order where to all appearances only chaos resides.

The splendor of their sensory life graces children with an accompanying gift, an excellent memory, for so present are they to the perceptual richness of everyday life that little is lost or disregarded and all is stored in their memory banks for later review and delectation. They also can learn to become time players, able to speed up subjective time when they need it to go faster or slow it down so as to savor lovely moments or have more time to rehearse skills or review projects.

Should their natural zippiness and boundless curiosity entice them into situations where they are physically hurt, children can be trained through biofeedback to control any bleeding and even to accelerate their own healing. We now enjoy a wealth of recovered knowledge of indigenous healing procedures; because they are innocent enough to accept them, children are naturals to learn these skills.

Moreover, like the yogi adepts in the Himalayas, children have the capacity to learn to voluntarily control involuntary physical processes—self-regulating skin temperature, blood flow, heart and pulse rate, gastric secretions, and brain waves. Indeed, they can learn to enter consciously into alpha and theta brain wave states for meditation and creative reverie, drop into delta whenever they want to go to sleep, and call on beta waves when they need to be alert and active. Scanning their body, they can self-correct any system whose function is less than optimal. In fact they can access the "optimal template" of their own physical pattern and learn to use this wisdom of the body for further self-correction and improvement.

As they develop appreciation and respect for their own bodies, children gain regard for the bodies of others as well. The moral consequences of this connection can be enormous: people are no longer seen to be abstractions or statistics but partake of an embodied glory that makes killing or violence unthinkable.

Consciousness Training. A Jump Time education would also require that children learn to self-orchestrate along the continuum of states of consciousness, traveling interior highways through realms of fantasy and imagination, spelunking through caves of creativity. We have discovered that consciousness has many states, apart from that half-awake state we call normal waking consciousness. Some states are hyperalert, allowing one acute focus and concentration. Others grant access to states of high creativity. And then there are those states in which the personal self seems to disappear and one enters mind at large—a unitive condition in which one discovers oneself to be the knower, the knowledge, and the known.

An arts-based education greatly facilitates the capacity to travel in inner space. Drama, music, and the potent richness of language teach children to think in inward imageries and to experience subjective realities as strikingly real. They become explorers in the vast reservoir of virtual realities without any machine to assist them. They discover with delight that self-creating works of art are always budding out of the fields of their minds and that they can capture and rework them as they wish. In essence, they are discovering one of the main secrets of being human: that we contain within us many cultures, many worlds.

Education in the Appropriate Use of Technology. Education these days goes hand-in-glove with technology. The challenge is to teach children to control the machines and not the other way around. So seductive is the new technology, so expert and computer literate are many children, that the computer is fast becoming an extension of their bodies. There they are, little protocyborgs, manipulating electrons before they can parse a sentence. Information is out there, a disembodied "fix" to be downloaded without much struggle by little posthumans with flying hands and screen-lit faces. Math is a matter of hitting buttons on the calculator, and art is moving preset graphics into place. The juicy world filled with the blood music of its people and their passions risks becoming an abstraction to be viewed from a space more suitable to the gods of Olympus than the life of an eleven-year-old. Like the mouse that eventually dies from pushing the pleasure button and forgetting to eat, the only life some children lead is on the screen. Thankfully, these children are still in the minority, but not for long. The revolution in high-tech expectation is forcing the hand of school boards everywhere, and computers in every classroom are becoming as ubiquitous as the proverbial chicken in every pot.

As much as I love computers and the technological utopia they portend, I for one am not interested in marrying my humanity to a machine. Many have suggested that in the years before 2050 we may have automatons that are conscious and self-aware. The MIT computer scientist Marvin Minsky believes that the next stage of human evolution will be the merger of humans with electronics. If he's right, our descendants might be able to transfer memories, thought patterns, even personalities clump by clump to neural nets contained within the electron circuits of a robot. An entire consciousness transferred, and the myth of the Golem is realized in a body of silicon and steel. That is why I believe that Jump Time holds the key to our future humanity: Do we become posthumans or deeply realized humans?

What schools can do to influence the outcome is increase children's experiences of high touch to balance the seductions of high tech. A high-touch education—holistic, integrated, arts-centered—calls forth the natural splendor inherent in every child's mind and body. Human beings contain far more images, ideas, stories, information, feelings, and, of course, consciousness than any computer. In a sense, we humans are metacomputers with the entire cosmos as our hard drive and our body-minds the screen for its unfolding. Western dualism, which has split mind from body, body from nature, and self from universe has tended to increase the chasm between what we think we are and what we really are. Thus, our dependence on machines for our reality. Thus, too, the importance of high-touch education to bridge the great divide and bring us home to the universe that resides within.

Properly balanced by high touch, computers can, without question, be adjuncts to self-discovery and exploration of the many worlds in which we live. Technology democratizes information and encourages the growth of noninstitutional, ever-shifting networks of self-organizing learners. Computers free students from the constraints of linear, word-based reports and allow them to express their understanding of a subject through multimedia creations, incorporating a rich composition of visual and auditory devices and providing pathways and links to other knowledge resources on the World Wide Web.

Schools can modulate technophilia by teaching children to use computers to enhance their experiences of reality rather than to substitute for it. A few snapshots from the frontier. A thirteen-year-old with cerebral palsy uses a computer to help track weather patterns and shares the results with meteorologists all over the world. Astronauts on the space shuttle and explorers in the rain forests of Peru relay the excitement of their discoveries as they happen to students all over the world via the Internet. Students at an Omaha school use the Internet to identify countries that are violating human rights, create multimedia projects, and send them to governments with pleas for reform. At Vermont's Cabot School, kindergartners through high school students take part in a School Report Night to which the entire community is invited. At one display, visitors click a mouse to view multimedia stories written and illustrated by primary students. At another, they watch a riveting documentary about the internment of Japanese-Americans at Manzanar Internment Camp created by a ninth-grader whose grandfather was among those imprisoned at the camp.

The humane use of human beings demands that embodiment be central to all educational experience, and that artificial intelligence, however fascinating, be our servant and not our surrogate.

Ethics and Values Education. Training in ethics and values is a road to freedom, for, as Benjamin Franklin wrote, "only a virtuous people are capable of freedom." Theodore Roosevelt concurred, claiming that "educating a person in mind but not in morals is to educate a menace to society." How we plant and tend a child's garden of virtues will have consequences for the future of the family, society, and the planet herself. For in these days, the garden comes preequipped with a very naughty snake who offers the fruit of many temptations and not a few paths to singular perditions. Adam and Eve had simple choices compared to the tangled undergrowth of our children's moral options.

It is a truism that moral education begins with creating a caring community within the classroom. A warm and supportive teacher-child relationship makes all the difference in the emotional climate of the classroom as well as in the cognitive development of the child. We all have memories of Mrs. Toad-Faced Horror who had us shaking in our shoes and peeing in our pants in expectation of chastisements that left us stupid and stunted. The chemistry of fear is one of the best ways to block the development of both heart and mind.

One well-known stratagem for creating a positive climate is encouraging children to discuss "the way we want our class to be," which opens into an exploration of the principles of fairness and kindness that make this goal achievable. Children themselves develop class rules that support these principles and are responsible for seeing that they function well. However, in Jump Time, more is required. Children can move from this discussion to a consideration of "the way we want our world to be." Classrooms can model a civilized society—a community of responsible, moral people who have zero tolerance for racism, sexism, violence, and psychological abuse. Cooperative learning, as when third-graders help first-graders with their reading, also encourages responsibility and consideration for others.

Enriched by a high-touch education, children are better equipped to appreciate values and make moral decisions. Being more, they come to feel and care more deeply about decay and degradation in the world around them. Storytelling drawn from folktale, literature, history, and biography is an important part of this training in moral wisdom. Children should be encouraged to dramatize traditional stories and to make up new ones that celebrate virtues. The enactment of an ideal is always one of the best ways to make it second nature. Storytelling can also help children develop an inner voice of conscience, one that can and will speak up for worthy behavior.

Another critical component in ethics training is understanding the consequences of actions, even establishing a Department of Consequences within the classroom. Here children can discuss and enact the short- and long-term results of behaviors they experience. Teaching children to understand how their behavior impacts the well-being of others encourages empathy and the development of "leaky margins" to others. In Jump Time this sense can easily be expanded to a feeling of kinship not only with other children but with all creatures and forms of Earth life. Older children can be provided with opportunities for social artistry—youth parliaments, for example, which establish the context and the training for future leaders. Parliament members can be observers and even participants in local and national governments.

Finally, we might consider as a radical Jump Time antidote to bored and disillusioned youth the alternative of some children finishing high school by age fourteen or fifteen. The next two or three years would be spent in service activities, with a kind of youth corps devoted to cleaning up inner city neighborhoods, helping out in hospitals and convalescent homes, improving parks and recreation areas, and coaching and mentoring younger children. Such training in active compassion could give young people hands-on experience of the world. It could also be an extended rite of passage, a transition from childhood into adulthood, that channels the energies and idealism of adolescence into activities that enhance moral development rather than prolonging a period in which too many young people find themselves in a rootless limbo. Deepened by experience and practical wisdom, seventeen- and eighteen-year-olds can go into jobs or on to higher education, bringing with them a breadth and depth of experiential learning.

Teaching-Learning Communities

Children aren't the only ones who need new skills for coping with Jump Time. Adult minds and spirits are stretched when we reacquaint ourselves with childlike perceptions and enthusiasm for learning. If we could experience in a decade of adulthood something approaching the mental growth that we make between the ages of three and thirteen, just think of how expanded our minds and spirits would be, how enlarged our perspectives, how wide and deep our knowings, how compassionate and searching our understandings! It is we adults, after all, and not our children who have savaged the world and perpetuated the malaise of war and wasteland. It is we adults and not our children who fail to participate sufficiently in the music of our minds to orchestrate the challenge of change of our Jump Time world. And it is we adults and not our children who through ignorance and narrow-mindedness bring technically complex short-term solutions to bear on many-layered social and economic problems and, so, fail repeatedly. Bridging the genius of childhood and its gift for learning with our adult responsibilities may save our planet. Becoming lifelong learners may be the key to our continuance as a species. Granted, there are many fine adult education courses and seminars, but so great is the need that something more is required.

Part of my work has been helping to create teaching-learning communities around the world. This has not been difficult, as movements currently exist in many countries to establish councils or circles for people to meet regularly and share their life experiences, their dreams, their processes of self-discovery, and their attempts to make a better society. A recent study funded by the Gallup Organization indicates that 40 percent of adult Americans are regular members of small voluntary groups that explore personal and social development. Among these are various support groups, religious study groups, dialogue groups, psychotherapy groups, Twelve Step groups, consciousness-raising circles, men's and women's groups, garden clubs, salons, ritual groups, and all the Moose, Elk, and their ilk. Sociologist Paul Ray made a further study showing that some 24 percent of the population, or 44 million adults, in the United States are part of a community he calls the cultural creatives, whose members constitute a rising "integral culture," with values based on personal development, spiritual transformation, ecological sustainability, social activism, and the honoring of the feminine. My experience of similar groups in Europe, Australia, and parts of South America and Asia suggests that this integral culture extends far beyond the borders of the United States. Taken together, what we might term the wisdom circle movement and the integral culture represent a significant manifestation of a unique and very necessary phenomena: the spread of noninstitutional growth communities devoted to developing human potential able to address the enormity of the Earth's problems. It would seem that our present Jump Time has loosed a movement in consciousness and community to bring people up to speed so that they become practitioners of the possible in an era of vast transition and social upheaval.

As I see them, teaching-learning communities must focus on a form of cocreative education in which participants meet regularly to stimulate, support, and evoke each other's highest sensory, physical, psychological, mythic, symbolic, and spiritual capacities. Groups may be made up of family, friends, colleagues, employees, students, clients, parishioners, neighbors, mothers, and so forth. Ideally, members include only those who freely choose to participate and who feel strongly motivated to do so. Groups can be as small as five and should be no larger that twenty-five. Generally, such communities are most open for growth when they are without a regular leader, with the function of guide or facilitator rotating among the members.

Working in groups helps eradicate one of the worst ills that afflicts Homo sapiens: the tyranny of the dominant perception. Having companions helps us to see that different beliefs and perspectives are enriching. As we recognize the enormous variety in ourselves and others, we stand in awe before the sheer abundance of human creativity, renewing our dedication to increasing the amplitude of body, mind, and spirit of each member of the community. Practicing growth processes in teaching-learning communities also helps us bypass another insidious human failing, the potential for sloth. Self-discipline and good intentions have a way of evaporating without some consistent external commitment.

Jump Time is the most critical time in Earth's history. Never has the ultimatum "grow or die" been a greater imperative. I believe that working in groups creates transformational synergy, that we can travel faster and deeper together than we can travel on our own. Through the evocation of one another, we expand the base of our concern, developing an enhanced relationship to our planet and intensifying our recognition of its needs as well as our willingness to respond creatively to those needs. Working in community, each person holds the dreams and excellence of everyone else in the group, so that should we descend into a period of depression or despair, our excellence and dreams are held by the group until we return to a healthier mindset. And if can do this for each other, we can hold as well our collective dream for the world's future. Thus, in working with a group, we should try to involve people who, in their faith in the future of humanity and the planet, are willing to work together with constancy and caring to develop and extend the presence of the sacred in daily life.

Most of the books I have written are designed with these kinds of groups in mind, providing experiential journeys into powerful adventures of the soul that are at once both universal and intensely personal. When we work with a great adventure like the Search for the Grail, with a mythic figure like Odysseus, or with a historical personage whose actions through time and legend have been rendered mythic like Gandhi or Cleopatra, we see the experience of our own life reflected and ennobled within the story of that great life and join our personal themes with those of universal reality. In stories of love and loss, death and rebirth, revenge and reconciliation, we meet ourselves writ large and gain a cache of experience that empowers us to act in the world in noble and creative ways. We engage together in processes inspired by these lives and legends and use state-of-the-art techniques to increase our sensory, psychological, and spiritual capacities. In so doing we gain something of the mental agility of Odysseus, the perceptual acuity of Emily Dickinson, the storying mind of Shakespeare, the spiritual exuberance of Hildegard of Bingen. Like Shakespeare and company, we become participants, actors, and playwrights in profound stories of growth, challenge, wounding, and transformation, thereby creating the conditions and impetus we need for extraordinary personal and communal growth.

I think Will would be pleased by our new adventures in education for young and old. He might even write a sonnet.

CHAPTER 4

HERE'S LOOKING AT YOU, KID

What follows is an adventure in the re-creation of ourselves: a way of using our relationships to reorder and expand mind, brain, and psyche. Revisioning our relationships is a form of conscious evolution through involution. By developing our awareness of the emergent forms of psyche and consciousness, we gain the capacity to remap the relational landscape of our world—person to person, culture to culture, nation to nation, people to planet. In this way, we lay the foundation for the *regenesis of society,* moving from an egocentric and ethnocentric to a global vision of both interpersonal and international relations.

In chapter 2 we looked at the protean, polyphrenic self as a key player in Jump Time. Now we bring the many selves we contain to the very act of looking, so that the well-tenanted realms that I contain meet and appreciate the many worlds in you. When we see each other in this way, old insular phobias we hold for those people, nations, cultures that we regard as "other" fade in the sun of complex perception.

Looking at *Casablanca*

"Here's looking at you, kid." Why does that line from the 1943 Oscar-winning classic movie *Casablanca* evoke such sweet satisfaction? Maybe because it is at once a celebration and a deep seeing, a near ritual statement of affirmation of the glory that one person can see and conceive in another. At the close of the film, when Rick sends his beloved Ilse off to live a nobler life with her husband Victor, a hero of the Resistance, he touches her cheek, speaks those words, and volumes pass between them. Though they will probably never see each other again, that moment is forever, and that is where they will abide. For in the intensity of the look, their lives are forever coupled in essence.

What does it mean to look at something—to really see it? John Ruskin has said, "The greatest thing a human soul does in this world is to see something. To see clearly is poetry, prophecy, and religion, all in one." To see clearly is to pick up a maple leaf on an autumn day and be so struck to the core with the intensity of veins shot through with gold that your mind and soul are gilded by the looking. It is to look into the face of a stranger and to know his life as a part of your own. It is to look at someone of a different age, culture, race, or social class with wonder and astonishment. It is to renew regularly seeing and appreciating one's life partner, family member, or long-time friend. It is to move beyond the familiar into the fascinating. It is to feel that there is always a richer, deeper story in the other, one that exceeds your expectations, transcends your categories.

I have learned not to let my concepts get in the way of my percepts, not to let my preconceived ideas block my ability to see and perceive afresh. This has been a worthy lesson, for I have found that no one I encounter is a simple "either/or" proposition; a person is always "both/and"—plus much, much more. If I find a person tiresome or disagreeable or catch myself going on automatic with someone because I am tired or distracted, I stop, reconstitute my attention, and listen to the other with fuller awareness. Those "stop" moments have taught me grace. The other person then becomes for me a revelation, a stand-in for divinity. The greatest teachings I have received have invariably come from those people with whom I canceled my automatic response and really took in the fullness of their presence.

One of the greatest problems with human consciousness is that we are our own blind spots and see ourselves, as well as others, through a glass darkly. Alan Watts once said that the paradox we experience when we try to understand consciousness is like the eyeball trying to see itself without a mirror or teeth trying to bite themselves. But I, for one, believe that there are ways of gaining access to levels of observation that fill in those blind spots and deepen both our seeing and our way of relating to the world and other people. It's all in the manner of looking.

If you were a higher sort of god whose business it was to pattern the beings under your charge with evolutionary capacities, how would you create an eye? Perhaps you would create it so that it looks both ways, like the Roman god Janus, who had two faces. Such an eye would allow you to see a doubled world, the world looking out and the world looking in. Then, perhaps, you would create inexplicable gaps between those worlds, a great parentheses, as it were, between the world out there and the world as it is perceived, so that your beings would fill this gap with their own treasures of imagination and experience. In this tension relationship takes place and pushes us toward growth.

The eye is a frame, a little picture window that delineates the boundaries of our reality. How you choose to see someone—to frame them—determines the whole play of emotional forces between the two of you and the nature of the relationship that unfolds. Moreover, how you take in the sense data of seeing and reimagine its forms determines how your life and your relationships proceed. Your inward seeing creates an environment as real and as influential—if not more so—than the tangible environment of your outward life. Thus seeing is a creative act in which we have enormous freedom, should we decide to use it, to revision our world.

The Four Levels

In my work I have noted and described four levels of the human psyche. I call these levels the *sensory,* the *psychological,* the *mythic or symbolic,* and the *integral or spiritual.* Each level has a unique style of imagery and content, a psychology, and even a metaphysics. In these levels we have within us layers of living cultures that have been accumulating, growing, and changing since we were born, and even, perhaps, before. If we are to widen our scope, as Jump Time demands, from the limitations of the egocentric viewpoint to the comprehensive vision of planetary consciousness, we must teach ourselves to see the world with fresh eyes on each of these levels. Then our personal relationships can be deepened and society can be regenerated.

On the sensory level, what we call vision is a grand, cocreative partnership between what the eyes see and what the brain reconstructs from memory and association. When we look at another, we engage, as William Blake suggested, in an act of constant creation, infinitely variable and renewable at every moment.

Psychologically, seeing involves storying. Some stories arise from one's personal memory banks. Some are pure creation, putting together people seen, places known, and experiences had. If novels did not exist, the brain would have to invent them to express the process of the storying mind. On the psychological level, images sponsor us and discipline us to create their possibility.

On the mythic level, one touches into the rich stores of archetypal tales and symbols. It is as if both you and I stand between two worlds, the personal-particular world of perception and memory and the personal-universal world with its broader contexts and more universal formulations. When I look at you from a mythic or symbolic place, I see you as Everyman and Everywoman. I may associate you with historical, legendary, or mythical beings and discover the broad patterns of your life as they relate to the figures of fairy tales, legends, and myths. It is then that I think, "I see you, Odysseus, hiding behind that funny mustache, those aviator glasses." Or: "Mary Magdalene, how's it going with your counseling clients? " Or even: "So Sir Percival, still wandering are we, still searching for that Grail?" When I perceive archetypal reality shining through another's everyday persona, my appreciation for them deepens, I call forth that depth of possibility in the other, and frequently—though not always—they rise to the occasion.

When we look at each other on the spiritual level, we see each other as gods in hiding, sometimes experiencing each other as waves of pure energy, transparent to transcendence, the life force in its infinite oscillations. In such seeing, one is able to see the other's possibilities with a natural felicity one had not known before. It is the entelechy of me knowing the entelechy of you—god-self to god-self. In such seeing, there is the coincidence of opposites, the cessation of all contendings, the seeing of all forms in one form.

With a world in the throes of whole system transition, working with these four levels in a more conscious way can yield capacities that override outdated and dangerous constructs. Doing so gives us an inward sufficiency complex enough to cope with the overload of Jump Time challenges.

Seeing on the Physical Level

On the physical level, how actually do you see? Start with the Janus eye itself, considering the modalities of both outward and inward looking. Until fairly recently, it was thought that light waves gathered information from objects, which then went rushing along the optic nerve into the brain. Not so; for not a speck of light reaches the brain from the retina. Rather, seeing is a complex interaction of retinal cells, cones and rods, ganglions, optic nerves, primary and higher visual cortical regions, and ions of visual data leaping chasms from synapse to synapse backed up by neurotransmitters. Brain mapping, a hot science in Jump Time, has shown us how dispersed the brain's visual field is. One part of the brain recognizes faces, another the expression on the face, another connects names with faces, while still another relates the face to all other faces you have seen. All this information—80 percent from many parts of the brain and only 20 percent directly from the eye—comes into headquarters, the lateral geniculate nucleus, where "seeing" occurs. The brain then feeds the information gathered back to the eye itself.

Look at an unfamiliar face, and your brain might tell your eyes, "Hey, pal, what you think you are seeing does not conform to your present brain map; better look again." We do and, more and more often in Jump Time, we see a person with a "world face"—a great melange of Caucasian, East Indian, American Indian, black, and maybe a soupçon of Malaysian. "You're right," we tell our brains. "This is new. Better add a new section to the brain map." William Blake, we discover, was right: What we see is built up from an infinite variety and supply of combinations coming from various fields.

Since seeing happens more in the brain than in the eye, to understand it we have to journey inward to the image-making process itself. If you close your eyes and press gently on the eyeballs—what do you see? Perhaps you see bits of line, arcs and curves. Then as you watch, higher visual constructs often begin to form: snowflake, spider web, and honeycomb patterns; parallel lines overlaid in grid forms with small spheres moving along each line; radial fanlike lines; checkerboard shapes; dots of dark and shadow; swirls or geometric patterns; something that looks like a time exposure of stars moving at night. These are the ideoretinal patterns, the universal form constants of the brain. Out of them our higher cortical structures build the rich imagery of the inner world that we project into the patterns we see in the outer world. As Joseph Chilton Pearce describes it, in speculations that are close to my own:

> These forms involve, and/or arise from, the frequency realm from which our visual world is created; neural fields within our brain draw on and combine with the infinite varieties of shapes available through these pattern fields. (*Evolution's End,* San Francisco: HarperSanFrancisco, 1993, p. 53)

What is important to remember is that these shapes are universal; every human being has them. They are the basis for the creation of all visual reality.

From these form constants, eidetic images emerge—images held, as we say, in the mind's eye. Some resemble prolonged afterimages; others result from memories or random imaginative scenes. If one is in a particularly receptive state, the brain puts on a show—sometimes even a song and dance. We recall the face of a particular person, and everything unfolds within us.

> The way you wear your hat,
> The way you sip your tea,
> The memory of all that,
> Oh, no, they can't take that away from me.

As the song says, the brain's deeper eidetic images can be scanned like a photograph for detail. The problem is that we are often haunted by recurrent eidetics, images of people with whom we have had some relationship.

The way your smile just beams,
The way you sing off-key,
The way you haunt my dreams,
Oh, no, they can't take that away from me.

In the song by the Gershwin brothers, at least, we have the power to work with eidetic images for the remembrance of pleasure and the transformation of pain. For, as we all know, inner images are often loaded with physical and psychological content, a pain in some part of the body, a wash of emotion or depression. Eidetic imagery therapists work with clients to change the nature of the imagery, move it about, and repattern its forms so that the toxic effect of the recurrent or stuck images can be released.

If you want to try this, image someone with whom you have had a difficult time. As you see that person in eidetic memory, chances are the image has negative content and an unpleasant feel. Now, place that person in another context altogether. See them making cigars in a banana republic; put the two of you together as clowns in the circus; or have a wild and woolly adventure together. Then return to your original image and see how the emotion surrounding it has changed. It is likely that when you next see this person with your physical eyes, your reconstituted imagery will sponsor a friendlier and more spontaneous relationship. When Hamlet said, "There's nothing good or bad but thinking makes it so," he was recalling the power of eidetic imagery to impact perception.

So seeing, on the physical level, is accessing a vast electrochemical complex involving billions of cells and multiple brain activities out of which arise our fears and joys of those we meet and our marching orders for how to proceed in relation to them.

Prejudice is the pathos of limited seeing, the damping-down of the brain's potential, a one-track catastrophe that objectifies the other as it subjugates the self. When we allow our seeing to be cut off at the level of superficial differences—skin color, body size, age, style of dress, social class—we lose something of the mystery and magic of the other and the magic and mystery of ourselves. I am less, if I do not see you as more. Living is an experience of mutuality; to exist is to be connected. From atoms to galaxies, the universe is a continuum of communion. For too long we have been trained for xenophobia, for seeing the other as alien. We would like to think that we have moved forward, that we gotten beyond such ancient prejudice. In many ways and in many places, we have. And yet, in Belfast, Protestants and Catholics are still divided by thick walls of suspicion. In India, growing fundamentalism on both sides keep Muslims and Hindus in a state of wary hostility. The Balkans and the new republics of the former Soviet Union are powder kegs of ethnic animosity. Even in America, the chasm between rich and poor, majority and minority still threatens to swallow us all. In the interconnected world of Jump Time, these attitudes limit our personal potential and threaten our collective existence.

Relating at the Psychological Level

How, then, can people cross the great divide of otherness, step out of stereotypes and into seeing and honoring each other deeply? This question takes us to the psychological level, the layer of mind in which story reigns supreme. Human connections are engendered in the mother ground of shared story. Consciously or unconsciously, we seek to relate to people whose stories run parallel to, advance, or complete our own. Story is powerful and primal. Whether we are entranced by the dancing flames of the communal campfire of some shamanic yesterday or by the flickering pixels of a television screen, stories bring us home to the hearth of our true nature. In story we are taught lessons that move us beyond the serial monotony of diminished existence. In story we come alive to the passion play that is universal in ourselves and others, and thus our compassion grows. Looking at the world psychologically, we unlock an awareness that everything and everyone else in this world is replete with story.

Relationships spark and crackle in the telling and the hearing of shared stories. For as the ancients knew and many scriptures testify, "In the beginning" God spoke, sputtered, sang, chanted, and the world came into being. As a result, all relationships are ultimately a dance of constantly changing energies, frequencies, vibrations. Set a string vibrating and listen as an entire scale is heard—not just a single note but all its overtones. Tell a good story, and discover that it contains all stories—not just in the actions recounted but, as in music, in the relationships between the notes, the silences, the rhythms. We are that string, and when we are plucked by the hand of story, we vibrate an entire scale.

Which brings us back to *Casablanca.*

Several years ago, when a poll was taken to determine the hundred best movies, number one was *Casablanca.* Why, out of all the great pictures ever made, did the film community pick this picture as the supreme example of filmmaking art?

Casablanca was filmed in chaos and confusion, written and rewritten by a committee who often had no idea where the plot and characters were going. The shooting script varied from day to day, even hour to hour. Poor Ingrid Bergman begged the writers and director to give her a few hints about what her character, Ilse Lund, was supposed to be feeling, whom she was supposed to be loving, and why. Humphrey Bogart, when he wasn't in a rage because of continuous brawling with his wife (who wrongly accused him of having an affair with Bergman), would just stand in the wings, smoking incessant cigarettes and shaking his head. Add to this a budget so small that a cardboard cutout was used to simulate the plane in the closing scene. The film was hastily released and expected to bomb. Yet out of mayhem came a masterpiece that owes much of its genius not only to the atmosphere of wartime adventure and intrigue but also to the rich complexity of relationships between the characters.

Think of *Casablanca* and immediately there comes to mind a man in a white dinner jacket: Rick Blaine, hard-boiled, cynical, self-centered. The love of his life, Ilse Lund, ran out on him in Paris the previous year as the Germans marched in to occupy the city, and he has pushed his feelings down into a bitter knot. When Ilse shows up at Rick's Casablanca café with her Resistance hero husband, Victor Laszlo, looking for exit visas to escape the Nazis, the plot thickens, relationships shift shape, new stories are told, and everything and everyone is revealed to be more complex and more interesting than we thought.

Casablanca depicts a world, like ours, in which everything is changing, everything is up for grabs. In such a world, human connections of all kinds assume supreme importance. Rick's longest relationship is with his wise and full-hearted black friend, the piano-playing Sam, who is never really told to "Play it again, Sam," though this is, perhaps, the most famous line associated with the movie. Sam knows Rick's vulnerability and tries to protect him. It's almost as if Sam holds Rick's soul for him until Rick is ready to reclaim it. One of the great advantages of friendship is that, in times of personal lostness and depression, our friends hold our essential character for us, giving us the confidence that what we once were and can be again is safely banked, gaining interest, while we wallow for a while in temporary troughs of self-noughting. Hints of what Rick once was are revealed through the valiant heroism and selflessness of Victor Laszlo, Ilse's husband, the perfect foil for Rick's withdrawn and whiny selfishness. As the story unfolds, we learn that, like Laszlo, Rick had been a freedom fighter, helping Ethiopia fight the Italian takeover and working for the Loyalists during the Spanish Civil War. Living now in moral limbo, he both profoundly admires and envies Victor, who represents the man he wishes he had the courage to be. Instead, Rick emulates the amoral cool of his other Casablanca friend, the sycophantic French police Captain Louis Renault, a small satyr of a man, with loyalties only to the prevailing wind, at this point blowing toward the German-collaborating Vichy government.

Both Rick and Victor have deep feelings for Ilse, Victor's wife and Rick's former lover, and she has equally complex feelings for both men. Victor, who has poured his passion into his political mission, comes across as an asexual saint, with a wife who is a helpmate more than a lover. Victor and Ilse met when she was young student; she fell in love with his ideals and married him to help in his cause. We note that throughout the film Victor gives her only a fatherly kiss on the cheek, and she treats him like an esteemed relative. This chaste behavior stands in stark contrast to her stormy and passionate encounters with Rick. At the time of their affair in Paris, Ilse had said nothing to Rick about her marriage, as Victor had been captured by the Germans and was presumed to be dead. So Ilse is caught between the two men in a counterpoint of passion and compassion, desire and dedication. And yet, in spite of their stylistic differences, Victor, the noble Czech patriot, and Rick, the scruffy New York street brawler, have each fought against tyranny. This essential likeness draws Ilse to both. As a young woman she fell in love with the nobility of one, but as a mature women she is transported by the erotic energy of the other.

Ilse's dilemma asks us to examine how different time zones in our lives demand different orders of relationship. Ilse and Rick's signature song, "As Time Goes By," celebrates the eternal truths that bind all lovers together.

> You must remember this
> A kiss is still a kiss
> A sigh is still a sigh
> The fundamental things apply
> As time goes by.
>
> And when two lovers woo
> They still say, "I love you"
> On that you can rely
> No matter what the future brings
> As time goes by.

But can we still rely on these fundamental truths? Or, rather, should we? It seems to me that in Jump Time, "I love you" is one of the most transitory of sentences, meaning one thing for one person or in one time zone and quite another for someone else or under other circumstances. When "I" make this stupendous declaration, which of my protean selves is speaking? And with everything changing, seemingly overnight, is the "I" who said "I love you" yesterday or last week or last year the same "I" who says it now? And if the meaning of "I" and "you" are deconstructing in Jump Time, what, then, can we say about "love"?

One of my favorite writers, Michael Ventura, love's genie in the bottle, has written:

> In love we reveal ourselves in each other's presence. . . . After the first flush of romance it will strip us down. Then love will call up everything within us that is not love, that it may be healed. So to stand in the state of love, and to remain there, is to be inundated and shaken by everything inside that is not love, everything drawn to the healing power of love. A power that does not let us go, does not let us off. It's dizzying. It's frightening. It has no rules. Its deep sweetness, and its great reserves of calm exist only in relation to its capacity for revelation. (*LA Village View,* February 11–17, 1994, p. 29)

No hearts and flowers here. For us, as it was for Rick and Ilse in Casablanca, love is the slayer and the healer. Falling in love, as we Jump Time folk do so many times over the course of our long and complex lives, even with the partner whose face we have seen over the breakfast table for thirty years, is an opportunity for revelation and for transformation. For couples who recommit their love and union on regular occasions, "I love you" can deepen and become luminous, as the I's multiply in full polyphrenic glory. And if all the you's do the same, the room is filled with a vast congregation of lovers!

In their first flush of romance, Rick and Ilse planned to escape by train on the day the Nazis marched into Paris. He waited for her at the crowded station, but she never arrived, sending him instead a tear-stained note of love and regret that says she can never see him again. He was left, as he says, "A guy standing on a station platform in the rain with a comical look on his face, because his insides had been kicked out." Eighteen months later, Rick is in Casablanca running the Café Americaine, an upscale nightspot and gambling den where people of all nationalities congregate. Since "everybody comes to Rick's," it is inevitable that Ilse and Victor show up looking for the contact who will sell them stolen transit passes so that they can fly to Lisbon and from there to America. In the shock of their reunion, Rick and Ilse play out a contretemps of anger and attraction.

Ilse keeps trying to get Rick to hear the true story of what happened in Paris:

> ILSE: Can I tell you a story, Rick?
> RICK: Does it got a wow finish?
> ILSE: I don't know the finish yet.
> RICK: Go on and tell it. Maybe one will come to you as you go along.

With tears in her eyes, Ilse tries to tell Rick about her idealistic, youthful marriage to Victor, but his anger and sarcasm blocks her from continuing.

> RICK: Yes, it's very pretty. I heard a story once. As a matter of fact, I've heard a lot of stories in my time. They went along with the sound of a tinny piano, playing in the parlor downstairs. "Mister, I met a man once when I was a kid," they'd always begin. Well, I guess neither one of our stories is very funny. Tell me, who was it you left me for? Was it Laszlo or were there others in between? Or aren't you the kind that tells?

With her story still untold and tears running down her cheeks, Ilse leaves Rick collapsed at a table, with his head in his hands.

This scene is a microcosm of the disease of projection, which so complicates relationships between individuals and between national, religious, and ethnic groups. When strong emotion is at play, empathy disintegrates. The old survival brain throws up barricades and concocts tales to justify personal and collective anger or sorrow. Rick is doing here what we all do—take our old stories and experiences and project them as expected outcomes, even though the situation in which we are now engaged may have a very different plot. In this way, we tragically assure that tomorrow will look exactly like yesterday.

The same kind of projection finds its way into the strife between nations. The feuding and fighting, ravaging and rapine have made this planet a killing field and a morass of anxious misgiving. In work that I have done in mediation and reconciliation in Northern Ireland, in India, and between European and indigenous groups in other parts of the world, I find that despite elaborate process procedures created by the best think tanks, or the most lofty and deliberate political and academic councils, old fears and distrust remain. Signatures may be affixed to agreements, but hearts have not been changed. Without storying, contending parties may agree to abstract principles, but there is no real meeting, no genuine human exchange. My technique is to persuade people to tell their most significant and heartfelt stories to each other, to meet at the level of deep listening before they get down to the "business at hand." When lives are shared, everything else follows.

I remember once in central India, under the auspices of the

Institute of Cultural Affairs, I arranged a meeting between executives of the Tata Corporation, a large international Indian firm, and a group of people from a village who needed better jobs. The high-caste executives could not believe that these low-caste villagers had the capacity to move into more challenging work and would not even consider giving them a chance to try. "They are ignorant, superstitious people," one formidable executive declared. "They are fit only to clean bathrooms and offices." The villagers held a few preconceptions of their own. "Those Brahmins do not even think we're human. How can we ever talk to them?" Here was not only prejudice but prejudice that had been rooted in the caste system for thousands of years. However, they agreed to a meeting in a temple complex.

After they had prayed together before a statue of Lakshmi, goddess of abundance, and the elephant-headed Ganesh, who removes obstacles, we began our conversation. It became a day of stories—often about the simple things in their lives, a favorite calf, a much-loved grandfather, what it is like to see the first greening shoots from the seeds one has planted, how difficult it is to learn to recite in Sanskrit. Somehow I managed to draw a parallel between the villager's greening shoots and the seed syllables of the executive's Sanskrit, and we all had a good laugh. One villager brought in his seven-year-old boy who was a genius on the drum known as the tabla and who beat out the most intricate rhythm, astonishing and delighting the executives. The little boy told us that he wanted to grow up be India's greatest percussionist. Then I asked both villagers and executives, "When you were his age, what did you want to grow up to be? Not just in your profession but your dreams of how life could be?" As executives and villagers shared their childhood dreams, differences blurred and similarities became apparent. Certain things are universal—love, children, yearning. And as they spoke together of these things, they began to lose their sense of "otherness"; and only then could they begin to speak of how both groups valued work in keeping with human aspirations. The result was that the corporation set up a training facility so that these men and women could learn the skills that would enable them to get the jobs they desired. Just as important, they forged new understandings across caste, across differences. As Rachel Naomi Remen reminds us in her wonderful book *Kitchen Table Wisdom,* when we see each other through the distorted lens of labels and stereotypes, "we are in relationship with our expectations and not with life itself" (New York: Riverhead Books, 1996, p. 66).

Storying is essential to this process. It allows the best of what we are to come together—reason and intuition, empathy and understanding. Story opens the window of direct perception and lets the clear light of human connection stream through. In *Casablanca,* Ilse persists until she too is able to reach across the barriers of projection and expectation and tell Rick her heart and her history. When Rick allows himself to listen, he discovers that Ilse left him in Paris not because she didn't love him but because word had reached her on the very day they were to leave that the husband she thought dead had escaped and was badly wounded, awaiting her arrival nearby to nurse him back to health. The cloud now lifted, Rick and Ilse renew their love, and she promises never to leave him again. Downstairs in the cafe, patrons are also coming together in a passionate song of renewed commitment. The German soldiers, led by the odious Major Strasser, have taken over Sam's piano and are singing a German martial song.

Victor defies them by getting the band to play a rousing rendition of the Marseillaise. Soon, everyone in the cafe is pouring their hearts into singing the song—the gypsy with her guitar, the barfly in her throaty voice, people from all nationalities rising to sing of the triumph of freedom over tyranny, liberty over oppression. In the wash of shared emotion—love, courage, compassion—differences melt, and human connections shine with fresh glory.

Rick now has a tough choice to make. He is hiding two stolen transit passes. Should he use them to escape Casablanca with Ilse? Give them to Victor and take his chances with Ilse in North Africa? Or follow some other plan? In the Jump Time of Casablanca, as in our own, all stereotypes break down in order that the deeper story may unfold. Victor the virtuous asks Rick to use the letters of transit to take Ilse away from Casablanca to safety. Rick, impressed by Laszlo's courage and unselfishness, breaks his precedents and orchestrates change and redemption at every level. In a supreme moment of going beyond his expected role and desires, Rick urges Ilse to leave Casablanca with her husband.

> RICK: If that plane leaves the ground and you're not with him, you'll regret it. Maybe not today, and maybe not tomorrow, but soon and for the rest of your life.
>
> ILSE: What about us?
>
> RICK: We'll always have Paris. We didn't have it. We'd lost it, until you came to Casablanca. We got it back last night.
>
> ILSE: When I said I would never leave you . . .
>
> RICK: Ilse, I'm no good at being noble, but it doesn't take much to see that the problems of three little people don't amount to a hill of beans in this crazy world. Someday you'll understand that. Not now. Here's looking at you, kid.

Rick has moved into mystery and a kind of sanctity. His story has now joined the larger story of a world in transition, and he exists to serve that story for its nobler ends. Captain Renault also undergoes a turnabout of the sort that we see in Shakespearean drama. He abandons the German-appeasing French police and leaves Casablanca for Brazzaville, where there is a Free French garrison, taking Rick with him. Rick's final words acknowledge that human connections are our lifeline when the world around us is dissolving into chaos: "Louis, I think this is the beginning of a beautiful friendship."

So out of the shifting, unveiling, dissembling and reassembling of the characters and their stories, and their deepened seeing of each other, comes a classic portrait of the ambiguities of the human heart in search of relationship. All the great psychological themes are present in *Casablanca*: love lost and found; betrayal and forgiveness; selfishness and selflessness; and the ways in which people are always more complex than anyone thought they were. Taken seriously, relationship hones us into our many dimensional selves, rich in stories, loved and loving, in transit from who we are to who we are becoming.

Relating Mythically

When we journey inward beyond the psychological level we find that consciousness is charged with archetypes and symbolic dramas, as old as the human race, as vast as culture itself. Beneath the storied psychological realm is the place of Great Story itself, the mythic patterns that have organized consciousness and creativity since time out of mind. I like to think that the twentieth century pursued the intensive study of myth and symbol so that the twenty-first might use this material to good advantage. Mythologist Joseph Campbell and psychologist Carl Jung were among the many who helped bring to life the power and importance of mythic themes—seeing them as organic constructs of the psyche, as patterns to instruct us, warn us, and lure us onto a greater journey. Myths themselves are cosmic codings as well as the DNA of the human psyche, the source patterns originating in the ground of our being. Mysterious agents that never were in time and space but nevertheless are always happening, they give us the key to our personal and historical existence.

My own work has shown that myths are modules of collective intelligence, open-ended blueprints of the journey that is ours. Whereas most creatures come in hardwired with instincts that guide their life stages, we arrive with nothing like their instinctual advantages, and we have to keep on a strenuous learning curve to make it through life's many mazeways. Over many years, I have guided thousands of research subjects and seminar participants into their own inner space and listened as they described what they found there: adventures of the soul so grand, so mythic, yet so filled with universal themes that I can readily testify to the existence of a collective pool of myth and archetype residing in each human being as part of their natural equipment. It would seem that myth and archetype hold a kind of metainstinctual code, giving us guidance and impetus as they illumine our transitions.

Wherever I am in the world, I find that working with a mythical figure or a historical person whose actions have become mythic gives people a sense of their own lives writ large. Archetypes are our great inward relatives. They hold the metagenetic patterns of our potential life journey. Seen in primal forms like the forces of nature, they are also found in personified qualities like Love, Beauty, Justice, and Mercy, but their most powerful expression is in patterns of relationship. When coded into myths, archetypes become our teachers in the many ways we relate to each other—to our families, our communities, and the divine spirit that inhabits us all. In the great myths of love and loss, such as the stories of Isis and Osiris, Psyche and Eros, Tristan and Isolde, we learn about the pathos and passion that perpetually confounds the relationship between the sexes. In the life story of the Persian mystic and teacher Jalaluddin Rumi, we explore the exquisite rapture that comes of awareness of the divinity within another. The Christ story teaches us to love our neighbor as ourselves, while Buddha's deeds show us compassion as the highest expression of human relationship.

When we come to appreciate the forms of these stories—or better still, as in my work and in the work of many psychologists and their ritual-creating counterparts in indigenous cultures, when we enact them outwardly in song, drama, and active process or in inward imaginal spaces—we begin to look at ourselves and at others in ways that exceed our local knowings, our limited history. I have seen a seventy-five-year-old woman become an avatar of Isis as she sought her lost beloved Osiris, in this case, her husband Sam, encased—not in a tamarisk tree, like Osiris—but in a Ohio nursing home with Alzheimer's. As Isis, Sarah grieved the loss of the Osiris/Sam with whom she had shared so many fruitful years. But then, following the pattern of the Egyptian myth, she discovered loving energy in the "remains" of Sam, as Isis found potency in the seemingly dead Osiris when she was able to conceive with him the child Horus. On the next day, when Sarah visited the nursing home, she told Sam stories of their life together, sang old songs, and played the scratchy phonograph records of their youth. Sam's eyes brightened and glimmered with recognition. For that afternoon at least, Sam and Sarah's marriage was resurrected, and they became the lovers they once were.

Having assumed the ancient stories and their personae, having walked in the shoes of folk who lived at their edges, we inherit a cache of experience that illumines and fortifies our own. In times past, mythically instructed communities—the Navajos, the Celts, the ancient Hebrews—tended to be more self-assured in their cultures because they had access to the rich storehouse of experience and tradition that myths contain. The sense of alienation that has grown steadily over the last several centuries derives in part from a loss of these potent stories that offered us an all-encompassing interpretive image of our relationship to the cosmos, to society, and to each other. Unstoried, we have gone spiritually naked into the world, replacing the lost meaning of our once powerful myths with fundamentalisms, totalitarian regimes, and a corporate consciousness that promised to return the lost harmonies we had known in myth but never could.

Now, in Jump Time, myths have jumped back from their long sleep in the collective unconscious and from those cultures where they have been preserved and celebrated. The world mind in its wondrous walkabout has brought these stories into public awareness so that today, myths are the stars in our imaginal firmament. They show up everywhere, not just in litany and liturgy but coming in through the back door of pop culture—in cartoons, comic books, computer games, science fiction, Disney animations. On any given night, the television screen blooms with mythic faces. Consider *Xena* and *Hercules,* great mishmashes of compressed myth, shameless when it comes to historical or legendary accuracy but roiling with mythic ebullience. The multigenerational *Star Trek* universe and the still unfolding *Star Wars* saga offer us regular glimpses into mythic worlds as scruffy and familiar as our own. My personal favorite is Wishbone, the small Jack Russell terrier who stars in his own television series, playing with great seriousness the central characters of the great myths: Odysseus, Sherlock Holmes, Faust, Robin Hood.

Spurred by technology and by the recognition that our everyday lives would seem legendary to all previous societies, we are in all practical ways beginning to live mythically. We regularly fly through the air, level cities with lightening strikes from on high, reverse the course of rivers, and communicate instantaneously with other mythic beings halfway around the world. At night we tune into omniscience, viewing on screen the troubles and triumphs of the gods and goddesses of Hollywood and Washington. With all this mythic abundance, we are now in the first stage of remythologizing our culture, gathering the raw materials and playing with them with intrepid amusement. The next stage, however, is the movement from play to incarnation—beyond the Cliffs Notes version of celebrity gods and television-series myths into relationship with the mythic soul of the world.

Not since times long past has the mythic sensibility been available to us, but today, with the help of science, it has returned, recast as modern quests for the meaning and origins of reality. In essence, our search outward into space and backward into time for the beginnings of the universe and for the ancestral inheritance of the human species is the current form of the mythic journey to find the primal parent. The bubble chamber of the nuclear physicist is a contemporary crucible, the linear descendant of the old alchemist's search for quintessence—the elementary particle of being. Now, like the mythic mage Hermes Trismegistus, we fathom the vastness in both directions, seeing ourselves as the fragile balancing point between macrocosm and microcosm.

Most myths end with a return to the home place, revisioned at the conclusion of an adventure-filled journey as a luminous kingdom. In Jump Time, home is Gaia as seen from space, the living Earth herself—home, very sweet home—a profoundly generative image that allows for the aperture of the psyche to open so that evolutionary impulses can come through. The vision of our fragile planet floating in space comes to us like the angels of old, bearing a message and urgently requiring a response. What does She ask of us? Nothing less, I think, than the birth of a new planetary psychology—universal, panpsychic, and transhistorical in scope. Archetypes and their impetus for crosscultural connections are the genomes of the symbolic psychology of Jump Time. The electronic nervous system of the global village has resulted in a spreading and sharing of the images and archetypes of societies radically different from each other in geography, history, and ethnicity. Myths and stories cooked in climates vastly different from ours are becoming the materials of our everyday consciousness. The rebirth of images and their crossfertilization through media and travel spurs us to rethink the archetypal possibilities of our human selves, now expanded to include all present knowledge and all previous visions of tribe, race, and social class.

One of the most formidable of these shared archetypal structures is the shadow and the ways it plays itself out in expressing the repressed and disacknowledged aspects of ourselves and our societies. More than therapy, more even than most peacemaking agendas, our shared apprehension of shadow archetypes from around the world—skeletal children in Africa, AIDS and other world plagues, the horrors of ethnic cleansing—offers new light on old problems. We waste so much of our conscious energy by seeing our shadows as entirely toxic, rarely allowing our vision to go deeper to see the incandescent energy for healing they contain. The shadow demands that we probe beneath the apparent hopelessness of world problems with uncommon vision. President Jimmy Carter, for instance, with a heartfelt genius that refuses to regard past patterns as obstacles to future breakthroughs, took to Camp David previously implacable enemies, Prime Minister Menachem Begin of Israel and President Anwar Sadat of Egypt, and, in the light of his high regard, inspired them to see the best in each other and to envision new hope for their collective future. Jimmy Carter and his wife Roslyn continue, post-presidency, to hold simultaneously the realism of the shadow and the potential for the light. With intellectual rigor and full-hearted seeing, they spur people on to nobility and transformation. They are among the best people on the planet, showing up whenever there is need and profoundly making a difference.

Moreover, archetypes now shared and interactive are themselves undergoing a whole system transition. In Jump Time, we are regrowing those psychological and spiritual potencies once known as gods, recharging them through their reunion with other versions of themselves from across time and space. When we hold the new archetypes emerging from this conjunction in an affective relationship—marry them, if you will, as Beloveds of the Soul—our human capacities are stepped up to mythic levels, even as the archetypes step down to partner our evolving humanity. This conjunction is the basis for world-making and the entry to new ways of being. For just as we are coded in our genes and entelechy for grand purposes in our lives, so archetypes may be similarly encoded with information and purpose to aid in the transformation of the planet toward its next stage of being. When we live in primary relationship to the archetypal, we gain access to the impetus and support of the source level of being, which urges us on through dreams, vision, and internal dialogue toward a deeper life and a richer service in our world.

Spiritual Relationship

At the fourth or spiritual level one discovers the most luminous of all relationships—the Divine-Human partnership. There is nothing new about this—mystics and poets seem always to have known that this connection is both the source and goal of the human journey. I am reminded of the Jewish mystical tradition, which asserts that God made human beings because he needed partners in creation. As the poet Rainer Maria Rilke wrote, "We are the bees of the invisible. We madly gather the honey of the visible to store it in the great golden hive of the invisible."

Perhaps the purpose of evolution is to grow us into cocreators who can play a conscious role in transforming the potentialities inherent in matter and ideas into new forms, better societies, richer meanings, high art. The archetype of Divine-Human partnership transmits to us evolutionary wave forms, impulses for personal and cultural development. When we come into phase with the creative energies of the universe, our scope enlarges such that everything we see—the little girl in the swing, the dog gnawing a bone, the old man in the nursing home, the snappy clerk at the checkout counter—becomes a celebration of the wild, exuberant, all-accomplishing energy of divine manifestation. In such partnership, we participate in the vigor and the generosity of Divine Life. The marks of this resonance are, according to the great student of mysticism Evelyn Underhill,

> (1) a complete absorption in the interests of the Infinite, under whatever mode it is apprehended by the self; (2) a consciousness of sharing Its strength, acting by Its authority, which results in a complete sense of freedom, an invulnerable serenity, and usually urges the self to some form of heroic effort or creative activity; (3) the establishment of the self as a "power for life," a centre of energy, an actual parent of spiritual vitality in other men. (*Mysticism,* New York: New American Library, 1974, p. 416)

The local personality takes on a wider consciousness, a deeper and more comprehensive universal self that amounts to a significant mutation of one's being. The seed within, which held and nurtured the divine spark, is grown into full maturity, and we are transplanted into the estate garden of Universal Life. Then everything is seen as full of light and meaning, everything falls into place in the great pattern of connections. Why? Because one is living both here and in another Order of Being in which the relationships *between* realities seem as important as the things in themselves. Perhaps that is why the person who reaches the state of mystical union—what Buddhists describe as One Taste—suddenly becomes so creative, so willing and able to do almost anything in service to all life.

Thus, what begins as an archetype of partnership blends into union by virtue of the deepening strata of the psyche. In mystical Sufism it is said that at this point one is digested and reformed by God. At the same time, in some sense, God becomes human for us, as in the mystery story of Christ's incarnation and death, a story that tells of both the epiphany and the great demand for continuing relationship with the Divine. Thus Meister Eckhart's powerful declaration that God says to us: "I became man for you. If you do not become God for me, you do me wrong." The requirement is radical; the consequence is the dissolution of all boundaries. Eckhart again: "If I am to know God directly, I must become completely God and God I; so that this God and this I become one I." We become part not just of a larger psychology of being—the story of the soul of the world—but of the secret life of God. Mystical union is not intellectual or speculative knowledge; it is entirely experiential. And yet someone like Eckhart, one of the most powerful conceptual and experiential mystical theologians of the Middle Ages, was clear in his statement that the mystical relationship is not the privilege of the few but the vocation of all of humanity. He says this because he believes that God is entirely immanent in us, is in fact our very being. The divine-human partnership, then, is the essence of who we are. The quality of our Selfhood depends on our recognition of the divine image in us, on the degree of our immanence in the archetype of God. Or as Eckhart says, "The eye by which I see God, is the same eye by which God sees me."

With this eye, then, nothing seems alien or unworthy of our love. Ordinary reality is unmasked, and one sees all beings in shining splendor. As the English mystic Thomas Traherne wrote when walking into a town holding the God's-eye view:

> The men! O what venerable and reverent creatures did the aged seem! . . . And young men glittering and sparkling angels, and maids strange seraphic pieces of life and beauty! . . . Boys and girls tumbling in the streets, and playing, were moving jewels. (*Centuries of Meditations,* Morehouse Publishing, 1986, iii, 3)

Here's really looking at You, kid!

CHAPTER 5

PEACEMAKING AND THE GLOBAL LONGHOUSE

Dig deep into the cultural soil of America and you find an ancient treasure, perhaps more valuable than the technology, wheat, and other riches America exports to the world. Long before the arrival of the European colonists, a complex, people-centered government existed on these shores, one that provided a model for both the American democratic experiment and for confederated governments around the world. In this time when the world so desperately needs patterns for peaceful and effective societies, this model rises again, providing a much-needed prototype for the genesis of new ways of being in community, given a global society.

I would like to retell the story of the founding of this government—the most important story I know. It is a legend about a bringer of peace, a very great hero, a creator of community, a changer of his world. Since it is one of the foundation stones of American democracy, I acknowledge with chagrin not only how little known its details are to non–Native Americans but also the history of oppression of the people whose story it is. I wrote a book about it, *Manual for the Peacemaker*, and have used the story many times in my work, in 1999 at a conference in Dharamsala, India, with a largely Tibetan audience for whom the story paralleled the strife that has gripped their homeland and, in Belfast, as the basis for rethinking the Politics of Reconciliation in my plenary speech at the State of the World Forum. Every time I tell it, people at every level of society react powerfully. Mysterious in its effects, potent in its pragmatic power, it calls the hearer to join those who were with those who will be in the *regenesis of society* and the remaking of the social order. A Jump Time from the past, it provides a model and a template for the creation of a possible society in which equity, sovereignty, and human dignity prevail.

The story tells of a man whose name is sacred to the Iroquois and not to be spoken aloud, and I take a risk in writing it: Deganawidah, known as the Man from the North, celebrated and honored even today as the Peacemaker, whose heroic efforts inspired a peaceful and prosperous society where previously there had been long years of violence and intertribal warfare. Though the incidents of the story are the stuff of myth, it is believed that Deganawidah actually lived. Sometime between the eleventh and the fifteenth centuries, it is thought, he and his allies created a dynamic democracy among five tribes of native peoples in the northeastern woodlands, one that lasted hundreds of years and is still, to a remarkable extent, in force today. European settlers gave this nation the name *Iroquois.* They call themselves the *Haudenosaunee,* "the People of the Longhouse." The five tribes were the Mohawks, the Seneca, the Cayuga, the Onondaga, and the Oneida. Later a sixth tribe, the Tuscarora, were added. As we cast about for designs for repatterning our local and global communities, the image of the Longhouse seems wonderfully apt. Given our world of so many countries and cultures, corporations and consortiums, we could do no better than adopt the Longhouse as our overarching metaphor.

Deganawidah was a new kind of Peacemaker, a new kind of human being. His peace embraced what he called the New Mind, a radical change in consciousness that opens itself to a new order of health, justice, and creative power. This new order is expressed through dynamic conversation, intensive sharing of ideas, councils, ceremony, and cooperation. The pattern laid down by the Iroquois Confederacy, with provisions for initiative, referendum, and recall, and for voting for women as well as for men, served as a template of possibility for a new kind of nation—smaller autonomous states within an embracing State, tribes whole unto themselves yet belonging to a confederation, with bonds of kinship stretching through the whole.

For us in Jump Time, the Iroquois model points toward a new type of dynamic society within a diverse world community, a confederacy that honors the authenticity and the autonomy of each individual, nation, state, tribe, or culture. It speaks radically to the open moment we live in today, a time in which anarchy, dissolution, and lack of guidance have wounded us almost to death. Yet this very woundedness, if we choose, can open us to one another and to our own deepest possibilities. Unlike many of our ancestors whose life path was laid out for them before they were born, we have the opportunity to create a new story for ourselves and our world. This task has made us mythic, for in Jump Time, we inhabit a twilight world between history and legend like that in which the Deganawidah story takes place.

The Story Begins

As do all great mythic tales, the story begins in mystery.

> A mother and daughter lived alone in a lodge on the north shore of what is now Lake Ontario. The daughter became miraculously pregnant, declaring to her mother that she had not known any man. Then, in a dream, an angelic presence, a courier from the Great Spirit, visited the mother and told her that the Great Spirit, the Master of Life, was sending a messenger who would bring word of a New Peace and New Mind. The women were charged to take good care of the child, for his mission would be to carry forth the sense of justice and a new life in peace for all the peoples of the world.

This opening seems familiar, of course, reminding us of the birth of Christ and many other holy children. Yet for those of us in Jump Time, it holds, perhaps, a more personal familiarity. How many of us have, perhaps miraculously, conceived ourselves to be part of a plan to help free the world from its pattern of war and destruction? Those of us who took our place among the angelic and bearded faces of the 1960s, girls with flowers in their hair, boys barefoot and beatific and making the peace sign or traveling in buses through the South sitting in against segregation, saw ourselves as messengers of a similar consciousness. We felt ourselves to be part of a great continuum of peace missionaries, for in spite of its wars, our century has also birthed among the greatest of these, from Gandhi to Nelson Mandela to Helen Keller to Martin Luther King, Jr. Like Deganawidah, they are evolutionary agents who enter into time with a plan for a dynamic new order.

The name Deganawidah means "The Master of Things." It also means, according to Iroquois elders, "He Who Thinks" and "The One with a Double Row of Teeth." The latter refers to the fact that Deganawidah was said to have stuttered. In many stories, inspired people who bring a new vision for humanity either come into the world with a handicap or soon acquire one. This makes heroes vulnerable and exposes them to ridicule. As a result, they are compelled to clarify their message and spirit so that they can blaze forth brilliantly in spite of physical limitations. In the twentieth century, peacemaker Helen Keller, in spite of her deafness and blindness, traveled the world fighting eloquently and courageously for the rights of women, the disabled, and the underprivileged.

> As Deganawidah grew, he began to talk of a new society based on trust, cooperation, and friendship. He offered [people] advice on ways to improve their lives and govern the tribe. His counsel was not always appreciated, coming as it did from a young person without office or authority. Early discouragement compelled Deganawidah to consider his message and mission more deeply. He began to wander deep into the forest, seeking his own New Mind.

How sad it is that the lives of our present political leaders are so complex and compacted that they rarely have time to stop and reflect before they take action. They substitute polls for spiritual practice and data for deepening. In a more perfect world, our leaders would be required to take regular "forest interludes"—time and place for contemplating the issues at hand as Deganawidah did.

> After many days of fasting and prayer, a vision came to him. Deganawidah saw the prodigious work that he was to do. He would go from person to person and from tribe to tribe persuading people to take up the New Mind and the ways of peace.
>
> He announced to his beloved mother and grandmother, "I shall now build my canoe. The time has come for me to set out on my mission in the world. Know that far away on many lakes and rivers I go seeking the council smoke of nations beyond this lake, holding my course toward the sunrise."
>
> The canoe Deganawidah built was made of stone. When he was mocked for this choice of materials, he said, "If my stone canoe will float, you will know that my words are true, my message real." And so he set sail in his stone canoe across the great lake.

Setting out on a mission across water in a stone canoe represents the willingness to undertake an impossible task in the service of an important truth. The truth Deganawidah carried could not be more important: that incessant warmaking is a crime against life. When a truth is important enough, no task is impossible. Doing the impossible hones our pluck and cunning and calls forth inner resources that we never knew we had. Given the need to achieve a goal, discover a cure, redress an old wrong, or create a new society, people will do the impossible. My old friend Margaret Mead regularly took on impossible tasks and performed them. She reminds us: "Never doubt that a small group of thoughtful committed citizens can change the world. Indeed it is the only thing that ever has."

Paul Loeb, in his inspiring new book *Soul of a Citizen,* tells many stories of citizens doing the impossible and living with conviction in a cynical time. He tells, for instance, of Baptist preacher Bill Cusak in a small town in South Carolina in the mid-1980s. He became concerned for his granddaughter's future in a world in which the nuclear arms race was accelerating. After reading a letter to the editor on the issue from a biologist at the local college, he sought the man out, and together they began to build a peace community from the ground up. They showed a video on the nuclear arms race to churches, civic groups, garden clubs—anyone who would have them—and crossed the racial threshold by enlisting an African-American minister in their cause.

Initially the peacemakers met with considerable resentment. Cusak had been a pillar of the local Rotary Club, but when he addressed them on the nuclear arms race, he says, "They kind of treated me like I had the plague." Eventually, however, Rotary Club members became more interested and began asking Bill for information on issues of war and peace. In the end, Cusak's efforts changed the town's culture, creating an atmosphere in which difficult social issues could be discussed with fairness and depth. Of his own transformation, Bill says, "It's been like a crowbar to my soul, cracking my personality, opening up depths to go beyond that superficial intellectual process of detachment. It laid a foundation for a style of life I'll be following for as long as I breathe" (Paul Rogat Loeb, *Soul of a Citizen,* New York: St. Martin's Press, 1999, pp. 185–6).

Like those peacemakers who have followed in his footsteps, Deganawidah left his old world and way of living behind, and

> traveled to a new one, moving always toward the sunrise. He arrived first at a settlement of warrior-refugees camped along a shore. They told him a terrible story of unending intertribal warfare, an endless litany of destruction, starvation, and death; of feuding factions, lawlessness, constant incitement to war, slaughter of innocent women and children, and cannibalism. They described a world stalked by assassins, where anarchy reigned, and people lived in a state of constant fear of reprisal. Fear bred weakness, for in this state there was little time or energy to direct one's thoughts to developing other areas of life. And out of weakness, men turned to war to assert what little power remained to them.

Turn on the news, read the papers, sign on-line, and you are flooded with an amplified, techno-fied version of the horrors Deganawidah confronted when he crossed the waters. From Bosnia to Belfast to Beirut to one's own backyard, one finds the same saga of a world gone mad. Is it brain-rooted, this mayhem, as many theorists think? Is the fight-or-flight reaction the remains of a savage past still moldering in our brain's backroom? Or are our egos so fragile that they require the bombast of belligerence to justify their existence? Is testosterone to blame, as certain feminists declare? Or is it ignorance, as the Buddhists say, which gives rise to hatred and anger, delusions that cloud the clear and luminous nature of mind?

Perhaps all these explanations hold a piece of the truth and testify to the fact that humankind is still the fetus of itself. It was not always thus. For most of our history as a species, we lived in small communities, with little warfare, shared intent, and a basic agreement about how to live together. But now we are in the great divide between tribe/nation and planetary community, and dread marks the crossing-point to another order of civilization.

As in all myths and sacred stories, a monster guards the threshold, and we will not be allowed to cross into the realm of amplified possibility until we fool its expectations and gain the allies, the intelligence, and the spiritual power to jump past the monster of hatred and fear.

Evolution gives us clues to what must happen. As we now know, evolution progresses through relationship and new connections, not through the harsh, alienated path of survival of the fittest. As the evolutionary record attests, it is connectivity and cooperation that succeed. Boundaries and borders are the place where cooperative relationships are forged, the very threshold of evolutionary life, but they are also the place where ancient fears and old insularities dig in to make their dying stand.

And yet, a curious phenomenon also operates at border crossings. During the Civil War, Northern and Southern troops sometimes camped across a narrow stream from each other. At night, when the day's fighting had ended, they would sing songs to each other: "Oh, Susannah," "My Old Kentucky Home," "Camp Town Races." Sometimes they even crossed the stream and shared bits of food and cherished passages from their letters, before regretfully saying good night and huddling in their blankets to await the howling of the dogs of war that dawn would bring. This pattern, repeated many times and in many different places, tells me something that evolutionary history has been announcing for hundreds of millions of years: what we are really looking at the boundary is relationship. The instinct for community is the juice of all becoming, and war is but the shadow of our yearning for connectivity. As I write, twenty-nine civil wars and other deadly engagements are raging around the planet. Some see this horror as the beginning of a firestorm that will consume the world. But, perhaps, as I hope, it is the last blaze of violence before the global entelechy pushes us toward a more inclusive and connected planetary family.

Righteousness, Health, and Spiritual Power

> For Deganawidah, crossing the watery boundary brought him to a place of new beginning. He stopped first at a house built by the side of the warpath. A woman of the Erie nation lived there. Her life consisted of feeding and caring for the warriors of any tribe as they traveled back and forth along the path to war. Deganawidah told her, "I carry the news that will end the wars between east and west."
>
> The woman was shocked. She had lived only in war and dread, knowing no other reality.
>
> Deganawidah pleaded with her, "Please don't serve the warriors, the raiders, the destroyers any more. The Great Peace will be served if you stop feeding the raiders."
>
> "But that's what I do," the woman protested. "That's my work. That's who I am. I am the woman who feeds the warriors."
>
> "Serve a higher cause," Deganawidah admonished her, "the cause of peace. Don't nourish the raiders. Then you will discover who you really are."
>
> In a hesitant voice that was nevertheless filled with quiet passion, Deganawidah spoke the words he had been considering for years, "I speak for the Master of Life, and the word that I bring is that all people shall love one another and live together in peace." He then laid out his vision for the new society that could bring about this peace. It had three equally vital aspects: Righteousness, Health, and Spiritual or Creative Power.

Deganawidah's words are extraordinary. Who and what are our raiders, anyway? And how do we go about not feeding them? On the simplest psychological level, our raiders are the toxic thoughts and conditions we live with and focus on, those negative habitual patterns that darken our joys and rob us of our peace of mind. The Peacemaker asks us to turn our thoughts in more positive directions so that we stop giving energy to these raiders of our inner peace.

On a much more challenging level, he asks us to stop, just stop, the processes of war-making, the machines, the killing contrivances, in order to serve a higher cause. He asks us to listen with the heart as well as the mind to our opponents, until we find a common story, a place of mutual agreement and shared humanity. He challenges our political leaders to find ways to negotiate without ultimatums and to stretch their conception of what is possible to make room for more comprehensive and humane solutions.

In persuading the Erie woman to give up her former livelihood and to work for peace, he offers her a new way to use her skills. She can continue to be a nourisher, but one who feeds the dream of peace rather than the raiders of death. Moreover, he holds out the vision of a society so vital in its programs, so visionary and yet so concrete in its details that it is a lure to the world's new becoming. Perhaps a sufficiently potent vision for the world might challenge us today, as individuals, nations, and corporations, to redirect our skills and the energy of enterprise into programs and policies that nourish the planet and its peoples, instead of feeding greed and paranoia.

Reconsidering righteousness, health, and spiritual power is very hard. It requires that we rethink just about everything. Righteousness implies that peoples and nations establish the custom and expectation of justice between and among themselves.

Deganawidah was proposing an ecology between cultures and peoples, a cooperative way of being in which no one is injured physically or psychologically and interactions between groups and individuals are cocreative and mutually nourishing. Health establishes a sense of balance in body and mind. It also incorporates the awareness of inner peace that comes when bodies are nurtured and minds are free and at ease, an abundance of vitality and well-being in which human capacities are developed to their fullest. Power implies "spiritual power," or *orenda,* in the Mohawk language, a kind of spiritual essence that unites all things. By tapping into the *orenda,* the creative force of life, one finds within oneself guidance and knowledge about appropriate laws and ways of being for a better society, a kind of spiritual politics in which true justice prevails. Moreover, by accessing this spiritual power, one also taps into a larger pattern in which one is able to perceive and to live out of the Greater Spirit rather than out of one's own lesser one. Thus, he invites us to embrace a New Mind, one that looks fearlessly at our present conditions and sees what is deeply needed.

Perhaps the best Jump Time example of Deganawidah's vision of a society based on righteousness, health, and spiritual power is the miraculous community of Gaviotas, established in 1971 by social visionary Paolo Lugari on twenty-five thousand acres of barren plains east of the Andes in Colombia. What began as a collection of researchers, students, and laborers committed to developing a new model for sustainable human community is now, some thirty years later, a compound of neat white cottages shaded by mango trees and bougainvillea, with guest houses, a meeting hall, a commissary, a school, a hospital, and about two hundred permanent residents.

The sense of balance Deganawidah spoke of as essential to communal health is symbolized in Gaviotas by the arrays of solar panels on the steeply pitched roofs of buildings, which heat water for the community using ingenious, locally developed technology, and by the garden of bright aluminum windmills dotting the landscape, also locally designed, which pump water for the village from deep underground. Nearly 70 percent of the food and energy needed to sustain Gaviotas is self-produced. Most electricity comes from a turbine powered by a stream; a hydroponic farm produces vegetables; and beef, pork, and fish are raised locally for private kitchens and for meals in the open-air communal dining hall. A model sixteen-bed hospital serves all who come without charge. Local Guahivo Indians, working with the National University's pharmacology department, are turning the greenhouse adjacent to the hospital into a laboratory for growing tropical medicinal plants.

Given that the physical needs of the community are met with balance and grace, the civic righteousness Deganawidah mandated is easier to achieve. Government in Gaviotas is by consensus, supported by a series of common-sense unwritten rules. To limit public disorder, alcohol is confined to private homes. To protect wildlife, guns are prohibited. Doors are not locked, and there is no need for police or a jail. People who violate the community's standards are ostracized until restitution is made for the infraction. Everyone works and is paid fairly for their labor, but meals, medical care, schools, and housing are free for all.

Spiritual power in Gaviotas is symbolized by the transformation of the land surrounding the village. After discovering that no indigenous tree could survive on the arid prairie, residents of Gaviotas imported and planted more than 1.6 million Caribbean pines around the village. The trees produce pine resin, which is sold for use in the manufacture of paint, turpentine, and paper. But the trees have yielded another benefit—a clear expression of the *orenda* of Gaviotas. Sheltered by the new pine forest, dormant seeds of native trees not seen on the arid plain for millennia have sprouted. Over time, Gaviotas's foresters plan to let this burgeoning native forest choke out the imported pines and return the land to its original state—an extension of the Amazonian rain forest. "Elsewhere they're tearing down the rain forest," Paolo Lugari told an interviewer. "We're putting it back." Already deer, anteaters, and other animals and birds are returning. In Gaviotas, it seems, Deganawidah's vision of a healthy, righteous, and spiritually self-sustaining community has taken root and flourished in the modern world.

Moreover, Gaviotas is no insular Shangri-la. The jewel in the crown of the growing eco-village movement, villagers envision seeding a network of similarly self-sustaining satellite communities in Colombia. They also recognize that outreach is part of their mission and are proud to serve as a model for similar communities in other parts of the world. If you can do it in Colombia, they say, there's hope people can do it anywhere.

> The Erie woman was moved by the vision that Deganawidah set forth, and she said to him: "That is indeed a good message. I take hold of it. I embrace it." Deganawidah then renamed the woman Jigonhsasee, the Mother of Nations, the Great Peacewoman. Together, she and the Peacemaker discussed ways his ideas might be translated into a matrix on which to build a new society.

The metaphor Deganawidah and Jigonhsasee chose for the new society was both homey and practical: the longhouse, the common style of housing among the eastern woodlands peoples of the time. It consisted of a framework of sapling elm poles covered over with seasoned elm bark, with a rounded door at each end. Sometimes as wide as 25 feet and as long as 125 feet, longhouses looked somewhat like an overturned canoe some 20 feet high. There were no windows, but in the high roof, smoke holes were left open, though they could be closed with sliding panels when the rains and snows poured down. From ten to twenty families lived in a single longhouse, each family with its own fire. The building had space along the side for food storage and utensils, as well as huge bearskin-covered bunks for family sleeping. From the rafters hung ears of corn, squash, drying apples, tobacco, roots, and tubers. A family's space was acknowledged as personal and could be sheltered from other families by hangings or curtains.

In Deganawidah's vision, the Five Nations would form a single Longhouse of nations, an extended polity in which each tribal group would hold its own place yet feel the power of the Confederacy stretching under and above them, linking them by clan ties and peaceful trade instead of constant warfare. An intricate system of relationships would provide for an intensive communal life. The eldest woman in the Longhouse would be the Mother. She and other women of the household "owned" the Longhouse. When a man married, he entered his wife's lineage and her family's dwelling place. Marriages were made between people of different clans; indeed, from outside any connection to the mother's blood kin. Children were born into the mother's clan. Clan membership transcended tribal boundaries; thus members of the Wolf Clan in the Onondaga nation had close connections to members of the Wolf Clan among the Oneida, the Mohawk, and others of the Five Nations.

This people-centered connectivity is not unlike the affinity groups now springing up around the world in Jump Time, their members united by a common social purpose, profession, or special interest that transcends national borders. What with the Internet and the variety and frequency of international conferences and gatherings, group members are often more connected to each other, despite being separated by thousands of miles, than they are to the people living next door. Like the clan structure of the Iroquois, these affinity groups are themselves meta-communities, linking new continents of ideas in the geography of the world mind.

Perhaps the most important reason why the Longhouse society worked so well was that women were acknowledged as equals in every way. The grandmothers were honored as the wise ones and were often the spiritual heads of the tribes. Women made proposals, often originating them, chose the male sachems (chiefs), guided and counseled them, and, when necessary, replaced them. Women acted as Peacemakers, and their clan connections with each other held the weave that kept the social fabric from unraveling.

Here, too, the Iroquois model speaks powerfully to our time. In the midst of so much change, the rise of women to full partnership with men in the whole human and Earth agenda is the development with the greatest social consequence. In a Longhouse of extended community, it does take a village—or a greatly expanded family—to raise a child, freeing women to bring their unique perspective to all fields of endeavor. Women hold the Mother-Mind—a mind that is not primarily linear, sequential, and objective but is rather circular, empathic, and narrative. In such a mind, solutions arise from an unfolding of levels of understanding, and the inward world has as much value as the outward. Because its approach is systemic rather than systematic, because it sees things in constellations rather than as discreet and disconnected facts, the feminine mindview is supremely concerned with the networking of the individual with the larger social organism.

Women's style of consciousness is well adapted to orchestrating the multiple variables of the Jump Time world and to getting along with its multicultural realities. As women move into leadership roles in many fields, they will bring an emphasis on the nurturing and cultivation of our whole being—even if it means turning our world upside down. I have seen discussions around campfires and boardroom tables from Africa to Australia, men in attendance throwing up their hands as women members insist that the process is as important as the product, that the relationship of the part to the whole is as important as the seeking of end goals. Jump Time is filled with examples of women leaders, rising to network, innovate, challenge, scold, and make things happen, in even in the most recalcitrant and patriarchal institutions. Yes, they get trashed. Yes, they become the screens for the most outrageous projections. But somehow, for the most part, they carry on, taking courage from the knowledge that they are part of the biggest change of all, from which there is no turning back.

Like the wise women of the Iroquois, today's women recognize that only in relationship—in a Longhouse of Longhouses—does the world have a future. Clearly transnational structures akin to the Longhouse are beginning to emerge, in the United Nations, the European Union, the World Bank, the World Wide Web, various international nongovernmental organizations, and other branches of the growing and complexifying world tree. Up to now, we have been addicted to our separations, believing that the divisions we feel between races, countries, religions, even between ourselves and Nature, cannot be bridged. Yet in these days of looming ecological catastrophe, the underlying commonality of our planetary future is inescapable. The worldwide linking of technology, economics, travel, and media is making clear that unilateral action is no longer possible and that no nation or group can take action without affecting everyone else on the planet.

But there is a deeper point. Scientists are now proving what mystics have long known—the cosmos is deeply and totally unified in some mysterious way. Things that seem to be separated are in fact connected in ways that transcend ordinary time and space. Immense levels of energy flow constantly through the universe, renewing each of us and everything around us at every instant. The universe, we are coming to understand, is a flow that arises moment by moment from the abyss of energetic nourishment in a process of continuous regeneration. This universe of unbroken wholeness is, in fact, a living Longhouse, a cosmic Haudenosaunee that continually re-creates itself. The task that faces us in Jump Time is not simply building the networks that make up the planetary Longhouse of nations and cultures. Rather, it's waking up to the fact that we and every living thing on this Earth already inhabit such a Longhouse. Becoming aware of our interdependence has never been more critical, because the species we call human is, at this very moment, in a race between consciousness and global catastrophe. In 1990 Václav Havel, then president of Czechoslovakia, electrified the U.S. Congress by saying:

> Without a global revolution in the sphere of human consciousness, nothing will change for the better. . . . And the catastrophe toward which this world is headed—ecological, social, demographic, or general breakdown of civilization—will be unavoidable.

Seeing the Cannibal's True Face

> Now Jigonhsasee asked Deganawidah, "Where will you take your message next?"
>
> "I go toward the sunrise. I go to the new," he replied.
>
> "That direction is dangerous," she told him.

The sunset stands for old ways and expectations. When we move toward the sunset, metaphorically we retreat into old habits, contracted attitudes, and tunnel vision. The horrors of the century now past exemplify what I call the sunset effect, a time when, like the sun in the western horizon, the old ways flare out with great authority and power before they sink beneath the horizon of history and culture. The sunrise, on the other hand, is always dangerous—unpredictable, precarious, full of ambiguities, intense light suddenly illumining dark shadows. In the case of Deganawidah, a particular danger lay along the path toward the sunrise that he had chosen.

> Jigonhsasee warned him, saying, "A cannibal lives on the path that leads to the sunrise."
>
> Undaunted, Deganawidah replied, "I am here to end such evils, so that all paths become open and safe for everyone."
>
> Deganawidah traveled to the lodge where the cannibal lived. He climbed up onto the low roof of the dwelling and looked down the smoke hole into a kettle of water on the fire.

> When the cannibal approached to stir the pot, his latest victim in hand, he saw a face reflected in the water and was stunned by it. Not realizing he was seeing Deganawidah's reflection, he thought he was seeing himself. And what a face! Strength, forbearance, character, even wisdom shone forth from the face he saw. "That's a great man looking out at me in the kettle. I had not realized I looked like that. That is not the face of a man who eats other people."
>
> Descending from the roof, Deganawidah embraced the cannibal and strengthened him in his resolve to transform his life.

The true Peacemaker is willing to encounter the worst of us, our shadow selves, our most fearsome nature. But he does so not by confrontation but by mirroring our basic goodness, our True Face. Whether in the media or across the international negotiating table, our task is cultivating the ability to look upon our potential enemies with respect. *Re-spect,* after all, means "to look again."

Too often, it seems, the media does just the opposite, cannibalizing people and their reputations for profit and ratings. The economics of infotainment dictates that the media make its living by creating fears and then providing the thrill of reading about these fears from a distance. Or if not fear, scandal. Or if not scandal, stories in which the shadow is amplified, regardless of how much light might also be there. Media has conditioned us to be titillated by the worst and bored by the best. Like the cannibal, numbed by the media's daily horror show of rape and rapine, massacres and drug wars, and the genocide of nations, we lose sight of the our own nobility. Switch channels, and more often than not, we are assaulted by dramatizations of the same stories, until we are no longer able to distinguish between cruel fiction and crueler fact. Our capacity for self-appreciation deadened, is it any wonder that we feel our humanity dwindle?

The good news is that the barrage of media negativity is, in many cases, having an unintended and opposite effect. The tragedy played out on the evening news may be bringing about the catharsis of all shadow, inspiring people to look deeper to find out what is really going on. In living rooms and chat rooms, business conferences and personal growth seminars, people are making an effort to meet each other at very deep levels. It is as if a universal yearning is rising for unmediated personal interaction, for the kind of respectful seeing that Hindus call *darshan,* in which one looks to see the god in others and to be seen oneself as a carrier of the divine. How might our public life improve, I wonder, if we afforded our political leaders the opportunity to be seen in this deep and respectful way? I have often sat in council with leaders of nations and corporations, shocked and saddened by the emotional toll of the daily media attacks they bear. Too often, the unrelenting wash of projections calcifies into a suit of psychic armor, hardening our public figures to both the world outside and the world within.

Rather than the public humiliation we seem to demand of our wrongdoers, through televised show trials or endless speculation on evils real or imagined on the talk shows, we might well consider how transgressors are dealt with among one of the peoples of South Africa. They place the wrongdoer in the center of the village and everyone sits around him. For three days and nights, everyone, young and old, tells the person what they appreciate about him. Nothing negative is said; only words of appreciation are spoken. Those who have witnessed this event tell me that as the hours go by, the scales of falsehood begin to fall from the person so spoken to until his true beauty shines forth and he is embraced once again as a beloved and valuable member of the community.

Now deeply seen, the cannibal, who had once been a great orator, told Deganawidah of the horrors that had driven him mad. He had lived among the Onondaga tribe, where he had spoken out against a terrible sorcerer of great power. The sorcerer, Tadodaho, had seven crooks of wickedness in his body and snakes in his matted and twisted hair. In retaliation, the sorcerer used his terrible magic to kill the orator's wife and four of his seven daughters. The orator had become so filled with agony that he took on the character of the sorcerer, saying, "I will outdo him in his evil." Having once been a great and good man, the cannibal was now filled with remorse at the terrible wrongs he had committed, and his misery was enormous.

Deganawidah tenderly consoled him, saying, "That is a wonderful story! Nothing is more difficult than breaking patterns set deep within us. A new kind of awareness, a New Mind, if you will, has been born in you today: a desire to see justice done; to restore health and sanity; and through this, a recognition of your path to spiritual powers.

"You are a fine speaker, with a beautiful and persuasive voice. My message requires just such a gift. I ask you to work with me. Together we can carry the Good News of peace and power to all people who can hear us."

"Your message is a good one," the orator replied. "I take hold of it. I embrace it. How shall we begin?"

Then the two sat down together and painstakingly worked out more of the details of the new story they would tell. Deganawidah then directed the former cannibal to return to his home tribe to win over Tadodaho. "Of course, he will drive you away over and over again," Deganawidah warned. "Expect that. But at last you will prevail. For this purpose, I will give you a new name. From this moment on you shall be called Hiawatha, which means 'He Who Combs,' for you shall comb the snakes out of Tadodaho's hair."

Hacking Down the Tree of War

After sending Hiawatha on his mission, Deganawidah took his message to the Mohawks. The Mohawk people found themselves much taken by the idea of the Good News, the High Peace, and the Longhouse of extended community with New Mind governing it. However, the Mohawk chiefs were suspicious. After all, the people's well-being lay in their hands, and one cannot just lay down arms and declare a peace. Or so they argued.

"It's not us. It's not us," they cried. "We could do it. But our neighbors couldn't! How could we dare to begin such a thing when our neighbors continue to attack us?"

To be the first to lay down arms and announce a policy of peace is a troubling act, especially in a world rife with justified paranoia. Most especially this is true if you are responsible for the safety and welfare of your people.

The Mohawk chiefs challenged Deganawidah, "We need a sign, some act that will prove you are not an enemy agent come to divide and destroy us—to make us weak and peaceful, so that your warriors can sack us."

Nearby was a deformed and twisted oak growing over the edge of a waterfall. "Climb into the tree that leans over the falls. We will cut the tree down, and you will fall with it. If you are still alive in the morning, we will become part of the Great Peace."

Deganawidah climbed into the ancient oak, and the people begin chopping at the tree with their hatchets. Deganawidah wrapped his legs around a branch, leaned down, and began to help hack down the tree with his own hatchet. A strong rain began to fall. The Mohawks stood back amazed as Deganawidah with a savage fury attacked the tree, which he named the Tree of War. Deganawidah said, "It is I who have said that the twisted Tree of War must be cut from your lives. It is I who will cut away this tree and fall with it into the gorge. No man should ever be afraid to cut falsehood from his life, even if it is the very thing upon which he is standing. For to stand upon a lie is to destroy your future peace and joy. There is no value to living if falsehoods are your roots."

When only a few more hacks were needed to fell the tree completely, Deganawidah stood in the branches and looked down at the now respectful faces of the Mohawks. The rain had ceased, and an extraordinary rainbow arched across the sky. Deganawidah pointed to the sky and spoke of the many circles—the sacred hoops in Nature, the cycling of the sun and the moon, the cycling of the seasons, the circles of our lives, and the Sacred Hoop of the Nations.

Then he pointed to the rainbow and said, "Each tribe is like one of the colors you see before you. Do you see where the colors touch and intermingle? This is the trade between tribes that live in peace. Where people honor the higher ways, beautiful colors come into being, and there is plenty for all. The rainbow is a half-circle that we must complete in our hearts. Except for self-defense, there is no place for violence in a heart that honors the teaching of the Great Spirit. Remove such error from your lives as I now sever this tree from its false roots."

With that, Deganawidah leaped higher into the tree and landed as hard as he could on a branch which arched over the cliff. With a mighty crack, the twisted oak split, and Deganawidah and the tree fell end over end the long, long way down to the ground.

The next day the Mohawks saw smoke coming up from an abandoned lodge at the bottom of the gorge. When they

> climbed down to investigate, there Deganawidah sat smoking his pipe. The Mohawks held a great feast for Deganawidah and embraced his ideas.

The twentieth century has been filled with people hacking at the tree of falsehood, knowing that the danger to themselves is enormous and that they might die in the gap that lies beneath them. Yet they persist in doing things that may seem impractical, impossible, and even absurd in the service of their truth. Mahatma Gandhi used truth force and passive resistance to win against guns and clubs, contemptuous colonial justice, and the crack regiments of the British Empire. Rosa Parks refused to acknowledge the falsehood of African Americans riding in the back of the bus and so launched the drive to desegregate America.

Sometimes falling into the gap can be a literal event, as happened to Carol McNulty. A teacher from Long Island, she was appalled when she saw a documentary, *Zoned for Slavery,* that showed a fifteen-year-old Salvadoran girl who for two years had been sewing clothes for the Gap clothing company, working eighteen-hour days and making fifty-six cents an hour. She was unable to attend school, take time for herself, or talk to other workers. Her beautiful brown eyes, hopeless and clouded with despair, mobilized Carol to take action. As Carol explained, "I saw such a look of helplessness in her eyes. . . . It's wrong for children to live like that—undernourished, without hope, literally chained to machines. She was just one young woman whose life was so blocked. If you multiply that by all the others, it's horrendous."

Carol and her husband Bill began weekly protests at a neighborhood Gap store. In winter rain or snow, they stood in front of the store, soon joined by a nun, some students, a few young mothers, a longshoreman, a union organizer, and others. They handed out literature, talked to the Gap's customers, and demanded an end to the terrible conditions and treatment of Central American workers. In so doing, they mended the broken link that makes people withdraw from civic life, splitting their lives from their values.

Of course, when one's political opinions become visible, one is exposed to abuse and paranoia. Carol and her group were reviled by many. The trustees of her village investigated ways of bringing antisoliciting laws to bear on the group. They even circulated the lie that the group was harassing people who tried to enter the store. When Bill asked the trustees to allow him to challenge this falsehood at their meeting, they refused, saying that they did not want to give him another platform for his protests. However, pressure continued to build, and more and more people began to picket Gap stores. Young Salvadoran women workers were sponsored to travel from city to city to tell their personal experiences of the company's practices to civic groups and churches. As a result of the ensuing outcry, the Gap agreed to change its policies and allow regular monitoring of its Central American factories by churches and human rights groups charged with making sure workers received better wages and treatment.

Carol and the other protesters celebrated their victory with signs thanking the Gap for doing the right thing. "Standing out there every week was hard," Carol said, "but it helped me to remember that the mothers in Central America love their children the same way I love mine. They bleed and feel pain like we do. We're not better than they just because we're more comfortable" (*Soul of a Citizen,* Paul Rogat Loeb, pp. 122–5).

Carol's words evoke what Deganawidah told the Mohawks

about the rainbow, in which the blending of tribes and cultures creates even more beautiful hues. Perhaps as much as trade and cultural fusion, the bleed-through of shared pain and compassion in a world of disparities between rich and poor, North and South, brings the sacred hoop of nations to completion in our hearts. Moving beyond compassion fatigue for the socially conscious and beyond the numbing effect of media news for social couch potatoes requires hacking down the tree of falsehood in very literal ways. Mind-numbing rhetoric and the safety of statistics are irrelevant in a world teetering on the edge of a gap.

A Ritual of Condolence

> While Deganawidah was bringing the Mohawks into the Longhouse, Hiawatha had returned to his tribe, bearing the Good News of the Great Peace. But the sorcerer Tadodaho threatened any who heeded Hiawatha and used his evil arts to bring death to Hiawatha's remaining daughters.
>
> Hiawatha's grief was inconsolable. Overwhelmed by the weight of personal tragedy, his mind became more and more distorted. Once again the evil soul of Tadodaho seemed to fill him like an incubus, twisting his spirit, scarring his mind, and making peace seem impossible.
>
> Only Nature offered Hiawatha consolation. As he wandered, Hiawatha picked up shells on the shore of a lake and made them into beads. When he came upon a stand of elderberry rushes, he cut the rushes into three lengths and strung the shell beads onto them.
>
> When he reached the edge of the forest near one of the Mohawk villages, Hiawatha sat down on the stump of a fallen tree — some say the same tree from which Deganawidah had fallen. Then he cut two forked sticks and planted them in the ground, placing a small pole between them. He [laid] the three

strings over the pole, saying to himself, "This would I do if I found anyone burdened with grief even as I am. I would console them, for they would be covered with night and wrapped in darkness. These strings of beads would become words of condolence with which I would address them."

That night, Deganawidah, who was living in the Mohawk village, went to the place where the smoke from Hiawatha's fire was seen rising. He saw Hiawatha meditating on the condolence strings hanging on the pole before him. Deganawidah lifted the strings from the pole one after the other and spoke the words of the Requickening Address, used even today by the Iroquois as part of the Ceremony of Condolence.

Presenting the first string, Deganawidah said, "When a person has suffered a great loss caused by death and is grieving, tears blind his eyes so that he cannot see. With these words I wipe away the tears from your face, using the white fawn skin of compassion so that now you may see clearly. I make it daylight for you. I beautify the sky."

Presenting the second string, Deganawidah said, "When a person has suffered a great loss caused by death and is grieving, there is an obstruction in his ears, and he cannot hear. With these words I remove the obstruction from your ears so that you may once again hear the sounds of laughter."

Presenting the third string, he said, "When a person has suffered a great loss caused by death and is grieving, his throat is stopped, and he cannot speak. With these words I remove the obstruction from your throat so that you may speak your truth."

By these ritual words, Hiawatha was freed from incessant dwelling on sorrow and grief and his mind was healed.

The strings of shell beads used in the Iroquois ceremony of condolence are called *wampum.* In Native American tradition, when you tell your wampum beads, you speak the truth of your pains, your path, and your stories. The Iroquois people say about the wampum, "this belt preserves my words." I have often used a version of this condolence ritual to help people lift their sorrows and deepen their stories, as well as stand witness to the sorrows and stories of others. The procedure is simple. Participants are invited to make a bead string, each bead representing a personal sorrow. Then, working in pairs like Deganawidah and Hiawatha, the partners take turns telling each other what each bead stands for—a person who has died, a lost relationship, even a dark quality of mind that they want to lift in themselves. For instance:

> The first one, this red bead, stands for my sense of deep loss about my husband, the twentieth anniversary of whose death I marked just last week. So he's really strong in my memory. He was a supportive, powerful, understanding male figure in my life, and I still feel his loss. And then I have these little white beads which are my in-between years which I spent doing nothing, working and trying to forget. And this brown bead is the farm that I grew up on, and the soil, the South Texas soil, that is part of my makeup and which I have left behind. And this green bead is the acceptance speech I'll never give for the Oscar I'll never win.

When one partner has finished speaking, the other takes the beads, and with great respect, touches them to the eyes, ears, and throat of person who has spoken, saying, "As the witness of your griefs, I hold your beads of memory, and I say: With these beads I lift the darkness from your eyes, so that you can see your losses with new vision. With these beads, I unblock your ears so that you can hear deeply and with new clarity your own words and those of others. And with these beads, I clear the obstruction from your throat so that you are capable of full expression and can speak powerfully of your sufferings and your joys. Now, please, tell me again what these beads mean." Having been deeply heard and consoled, the first speaker might say:

> This red bead is the passionate laughter of life that my late husband had and that I still hear. And it also stands for not just the love that I felt for him, but the love he felt for me. So I feel that love burn red again. And the little white years since his death are turned now to peace. And the soil of South Texas is the root of my being that I no longer deny. And the Oscar speech I'll never give is transformed to thanks to everybody for everything. (Jean Houston, *Manual for the Peacemaker*, Wheaton, IL: Quest Books, 1995, pp. 108–10)

After one partner has told the beads in this manner, the other member of the pair does the same and receives condolence. If there are others in the room doing the same process, people can tell their beads to other partners, new insights appearing and the story deepening with each retelling. Even more important, sharing grief in a communal ritual leads to the realization that loss is a universal human experience and thus to a deeper appreciation and compassion for our collective sufferings. Some of my students have used this ritual at memorial services for family and friends. While the ritual may begin with tearful statements of what each speaker will miss about the person who is gone, these generally give way to statements of appreciation and celebration for the life now ended and leave participants with a comforting sense of acceptance of the seasons and boundaries of life.

Of course the beads can also be used in communal rituals to grieve sorrowful situations in the world.

> This red bead represents the carnage and violence in Kosovo. This yellow bead is for the killing of the students and teacher

> in Littleton, Colorado. This blue bead is for children who are beaten and abused. And this gold bead represents my need to make a difference in all of this and my sorrow that I don't know where to begin.

Rituals are ancient forms that have always been used to illumine human transitions. The word *ritual* comes from the Sanskrit word *rita,* which comes from the same root as the words for "art" and "order." Ritual art, like the movements of sacred dance, creates organic order, a pattern of dynamic expression through which the energy of a tragic event can flow in an inevitable evolution toward a larger meaning. I believe that we might go a long way toward healing the traumas of psyche and history, our collective sorrows, and even the breakdown of our societies through public experiences that could best be described as therapeutic mysteries—rituals of condolence, reconciliation, forgiveness, and new growth. It is important that we not dismiss ritual and mystery as atavisms we have outgrown in favor of more "scientific" methodologies. Rituals are older than antiquity. Ingrained in the wisdom of the self and the galaxy, their symbolic coding unlocks our latencies and grants us restoration into the human family and the council of all beings. That is why public occasions of shared grief and condolence, times when we can weep together and tell our stories, can heal not only our own traumas but the hurts of nations as well.

A Ground Swell for Peace

> Deganawidah and Hiawatha set forth again as partners, visiting over the next years the Oneidas, the Onondagas, the Cayugas, and the Senecas. Through persistent, person-to-per-

son, village-to-village, tribe-to-tribe human contact, they slowly but surely started a ground swell for peace moving through the land.

A similar person-to-person technique is being used in Israel to help bring together Palestinian Arabs and Israeli Jews. Started by American peace missionary Leah Green, the Compassionate Listening Project brings groups of American Jews into Israel to meet with people, listen to their stories, and ask questions about their lives. The goal of the project is to give people living in conflict the experience of being heard without judgment and with a focus on the human values and reasons behind their opinions. Without such experience, Green says, dialogue meetings between Israeli Jews and Palestinians would often consist of canned speeches and people yelling at each other rather than healthy and productive conversation. Now Green meets with each side separately in a Compassionate Listening session before they are brought together to talk.

On a recent trip, people from the Compassionate Listening Project met with a group of Israelis, including many who had settled on the contested West Bank. With their very survival threatened on a daily basis, settlers are often hostile to Jewish peace activists. On this occasion, a woman settler was asked why she wanted to live on the West Bank. The woman spoke about her mother who had survived the Holocaust and walked across several countries to bring her children to Palestine. Responding to the woman's story, Compassionate Listening participants reflected back to her what they had heard. "What is really significant for me in listening to you is that I hear your incredible love for this land," the woman was told. Touched by having been heard so well, the woman broke down and cried. With this bond established, the woman was invited to meet with Palestinians who shared her love for the land.

How can we create a sustainable peace that can replace years of hostility and conflict? For Green, as for Deganawidah, the only way is the hard work of "meeting one's enemy and coming to know the human being behind the stereotype . . . of acknowledging the suffering in each other's hearts." Peace, Green says, grows "eyeball to eyeball."

> At last the chiefs of the five nations stood ready to join the new Confederacy. Now was the time for Deganawidah and Hiawatha to return to Onondaga Lake where the one dissenting chief, Tadodaho, the evil sorcerer, dwelled.

Tadodaho stands for everything that obstructs the way of peace: an energy that bears a tremendous negative charisma; a static and unmoving power; autocracy, including the inner autocracy of the ego; unwillingness to change; hunger for power and control; and the deep wounding that leads a person to wound the rest of the world. The twentieth century has seen too many modern Tadodahos. From Bosnia to Libya to Cambodia to the Third Reich, power-hungry dictators have brought millions under their merciless sway, making the century now past a horror and a holocaust.

We have always believed that it is necessary to destroy an enemy in order to win our goals. The Deganawidah story teaches us that so-called enemies cannot be destroyed; they must be transformed. How Deganawidah and Hiawatha deal with Tadodaho is a trenchant allegory of what we might do when we are similarly blocked by apparently implacable forces.

> Deganawidah and Hiawatha paddled their canoe toward Tadodaho's camp across the lake. When they were halfway across, they heard his voice keening a fearsome chant, "Asonke-ne-e-e-e-e-eh?" ("Is it not yet?")
>
> "Ah," joked Hiawatha, "he is impatient for our message."
>
> Almost immediately raging winds and waves, sent by Tadodaho as if in answer to Deganawidah's questioning cry, began to buffet the canoe. The wild elements were accompanied by the wizard's screaming declaration, "Asonke-ne-e-e-e-e!" ("No, It is not yet!")
>
> But Deganawidah and Hiawatha, knowing that the open moment was at hand, continued to paddle vigorously.

When you set out to make significant changes, know that, quite often, the waves of recalcitrance will rise to impede your progress. As the story tells us, you must not lose heart but renew your efforts and keep on going. Moreover, you can draw strength from the fact that whatever force is opposing you contains within it the energy that can be redirected to bring about transformation. In his cries across the water, Tadodaho was both denying the possibility of change ("It is not yet") and, at the same time, revealing that he was impatient for change to happen ("Is it not yet?"). Torn between hope and fear, as soon as he expressed his impatience for transformation, his mind slammed down, and he answered his own question from the place of stuckness. All over the world, people as well as institutions are in similar states of alternating hope and fear, wishing for change yet denying its possibility. This story tells us there is only one thing to do in such cases: maintain a steadfastness of commitment and break through entropy by offering many surprises, unexpected ways of viewing a problematic situation.

> Fearlessly, Deganawidah and Hiawatha approached the wizard. Deganawidah soothed him by massaging his body with sacred herbs in a holy medicine ceremony.

> Then Hiawatha spoke, saying, "These are the words of the Great Law. On these words we shall build the House of Peace, the Longhouse, with five fires that is yet one household. These are the words of righteousness and health and power. These are the words of the renewal for ourselves and our society."
>
> For a moment Tadodaho was drawn to the vision, but then he said, "No! No! What is this nonsense about houses and righteousness and health and power?"
>
> Deganawidah responded, "The words we bring constitute the New Mind. There shall be righteousness when people desire justice; health when people obey cooperative reason; and power when people accept the Great Peace."
>
> Again Tadodaho was drawn to the vision, but then he retreated to self-aggrandizement and said, "What is that to me?"
>
> Deganawidah replied, "You shall be of higher usefulness in this world. You yourself shall tend the council fire of the Five Nations, the fire that never dies. The smoke of the fire which you are tending shall reach the sky and be seen by all people."

It was not enough to persuade Tadodaho of the attractiveness of the Great Law of Peace; it was not enough to pacify him with good medicine; it was not appropriate to "bring him to justice" or "punish him." No, Tadodaho had great skills, enormous energy. Deganawidah offered Tadodaho the opportunity to use his tremendous talents for a higher good.

In a similar way, General Douglas MacArthur after the end of World War II treated the defeated Japanese as honored friends and allies, helping them rebuild their economy while empowering their self-esteem. As a result, Japan became an important partner to the United States and a creative and vital force in the world. The current attempt to use trade and education to bring renegade nations into the world fold speaks to the efficacy of the same policy. When Tadodaho-type people or nations are isolated and objectified, they cannot help but live out the negative expectations that are projected upon them.

> Then Deganawidah took Tadodaho before the assembled chiefs of the Five Nations and said to him, "Behold, my friend! Here are the Five Nations; their strength is greater than your strength, but their voice can be your voice when you speak in council. This shall be your strength in the future—not sorcery, but the will and creativity of a united people."
>
> Tadodaho finally broke his silence, and he said, "It is well; I now truly confirm and accept your message."
>
> Thus the mind of Tadodaho was at last made straight. Then Hiawatha, "he who combs," combed the mattings, the twistings, the snakes of distorted thinking out of the wizard's hair, and Deganawidah rubbed his body with wampum and herbs, with knowledge and with love. Then Jigonhsasee spoke to him privately, and some say mothered him, giving Tadodaho the nurturing and primal healing he sorely needed. As Deganawidah massaged him, he straightened the seven crooks of wickedness in Tadodaho's body that had filled him with hatred.

Deganawidah's battle for peace was nearly won. Now the task was maintaining the peace and governing the society that would enjoy it. To this end, Deganawidah devised a series of unforgettable symbols, potent images that would act to continuously keep peace alive in the minds and hearts of the people. The most powerful symbols involved trees.

> Uprooting a dead tree, Deganawidah invited all the warriors to bury their hatchets beneath it. Over the hatchets, Deganawidah planted a white pine, which he called the Great

> Tree of Peace. Its four roots of truth would spread in the four directions, carrying news of the Peace to all parts of the world. Any nation or people yearning for peace, consolation, and kinship could trace the roots back to the source tree itself.
>
> Deganawidah then spoke to the five tribes now assembled. "Greet eagerly the stranger at the door," he told them, knowing that a gracious and generous hospitality makes for a gracious and hospitable world. "Let old insults and wrongs be forgotten," he continued. Peace cannot flourish when we constantly rehearse our resentments. "Hunting and fishing grounds will be shared equally," he counseled, as the tribes now shared a single Longhouse and made up one family.

The Great Council

Perhaps the most important expression of the New Mind that Deganawidah mandated for the Confederacy was the council model of governance. He taught that social decisions must be made in councils that engage dreams as well as discussion, vision as well as practical programs. He believed that the dynamic patterns necessary to bring individual and universal good into every aspect of life and society are held in the Creative Mind of Being. Since all men and women are part of creation, these patterns are also contained in our minds. Open-minded discussion and mutual exploration in council, he taught, spurred inspiration and our discovery of these patterns.

> In the Great Council that Deganawidah established, representatives from each tribe would meet around the fire, kept by Tadodaho. The Council would meet at least once a year. Clear safeguards were placed on the leaders of the Council. Their job was to be generous, long-suffering, and sensitive, except in one

> way. Those that would be Chiefs, Deganawidah said, should have "skin seven thumbs thick so that no outrageous criticism could pierce them." Leaders were enjoined to maintain their center of truth, pray constantly to the Great Spirit for courage and clarity, and think always of the well-being of the people.
>
> Deganawidah also provided the nation with a pattern for meetings of the Great Council. Each opened with a prayer of thanksgiving to the Earth and to all that was in it. "I thank you for the Earth. I thank you for the waters. I thank you for the corn and for the harvest" This exhilaration of thanksgiving would warm the people's hearts with gratitude, making problems easier to address. And there were to be no short-term solutions. Among the Council's rules was that no decision was to be made without considering its effect unto the seventh generation to come.

How can Deganawidah's vision of the Great Council help us solve the problems that confront us in Jump Time? Let's imagine that you who are reading this book are taking part in such a council now. As we move toward a planetary civilization, it is essential that we send the white roots of peace all over the world. Thinking of yourselves as the gardeners of this world tree, what would you contribute to a discussion of righteousness, health, and spiritual power in our time? What would you say in council, for example, about these issues?

- How can nations preserve their unique cultural styles as we move in the coming century toward a world federation—a Federal Republic of the Earth?
- How can we renew the United Nations, so that it becomes more democratic, its decisions not shackled by a single nation's veto in the Security Council, which is tied to the results of World War II?

- How do we better manage world resources to preserve our global ecological heritage?
- How do we go about abolishing nuclear weapons worldwide?
- How do we shift the logic of conflict so that we understand that aggressors no longer win and that in any war both sides lose—an eye for an eye, and both go blind?
- Given Earth's fifteen thousand ethnic groups and six billion people, how do we make human rights a way a life so that we encourage the reunion of the human family?
- How do we empower emerging leaders who offer us the gift of hope?

Acting in extended councils meeting in town halls, living rooms, church basements, and on the Internet, let us allow Deganawidah's message to charge us to form a People's Movement toward building a passionate and sustainable future.

CHAPTER 6

WOK AND ROLL IN THE RAINBOW WORLD

A new species is being cooked in the cauldron of fusion. The membrane between cultures, between worlds, between old and new ways of being is breaking down. In the past, migrations and diffusions allowed for gradual changes and exchanges between cultures and identities. Now nothing is gradual; we are watching a speeded-up movie of strange multicultural mitoses and their stranger spawn. Cultures thousands of miles apart that had gestated in the womb of preparatory time for thousands of years are suddenly cheek by jowl, attending the same schools, working in the same businesses, sharing the same space, and, inevitably, bleeding into each other, sometimes in fury, sometimes in friendship. In the force of the meeting, a new genesis is occurring, occasionally a melding of genes but more often a mingling of previously divided worlds that, thrown together, undergo a sea change into something rich and strange.

What results is not merely a hyphenated amalgam, Afro-Asian rock music or Chicano-cyberpunk art but rather hybrid sounds for hybrid selves, a malleable, syncretic fusion that is generating its own cultural matrix. For human beings, the complexity of this not-yet-definable culture is providing sufficient stimulus to call forth latencies in the human brain-mind system that were never needed before, like bacteria learning to breathe rather than die when the culture of oxygen came in or, closer to home, the ways children absorb the mysteries of computer wizardry with ease while their parents struggle to master the basics. To see this cross-cultural stimulation in action, one has only to watch Western bodies pour themselves into Eastern yogas and martial arts, doing things that people bred on milk were never meant to do, given the calcium deposits on their knees. Or attend a workshop in gospel singing for Japanese tourists held at the Memorial Baptist Church in Harlem and watch people who had known centuries of bowing ceremoniously clap and sway from side to side and throats prepared for every formal courtesy belt out spirit-quaking songs.

The *breakdown of the membrane* I am describing is not merely cultural fusion; it is the joining together of geographies of the mind and body that have never touched before, weaving synapses and sensibilities to create people who are fused into the world mind with its unlimited treasures, its empowering capacities. This is what's happening in Jump Time: evolution as evocation, the quickening charge of cultural mitosis in our very cells. Through commerce and travel, technology and media, we are being catalyzed into ways of being that a few short years ago were the stuff of fantasy and myth. The shift in consciousness happening all around us is a kind of unstoppable if positive plague, a metavirus multiplying in our midst. From food to music to literature and theater—the very lineaments of culture—consciousness is remaking itself.

Fusion Food

Let's look at a few examples of this phenomenon, starting with food, perhaps the most basic and understandable component the of the fusionary world. After all, the palate is a palace of culture. The topography of the tongue alone travels the vast geographies of taste—sweet, sour, salty, spicy, bitter—each its own domain in the taste buds. Different cultures get contracted to specific alliances across these geographies, which is why Chinese food does not taste like Mexican, or French like Egyptian. Each nation of taste cultivates a distinct gustatory coalition of foods and their preparations that affects consciousness in very real ways, making for significant differences in people.

In the rapidly shrinking world of Jump Time, however, there are few countries where the main and side streets of the cities are not offering an embarrassment of world culinary riches. Take Melbourne, Australia, for example. Recently, I sauntered down Fitzroy Street; within a very short distance, I encountered, in this former outpost of overcooked sheep, restaurants offering the cuisines of Afghanistan, Ethiopia, Guam, Fiji, western China, northeastern China, South India, Belgium, Sicily, the

Philippines, Greece, Malaysia, Burgundy, Provence, Argentina, Tibet, and Tasmania. A sampling of these restaurants leaves the tongue tingling with a gustatory map of the world. And what it does to consciousness is provide the stimulants for a multicultural awareness, the world mind at table. Dine globally, think globally, might be the new maxim, the secret ingredient of international peace. Tell me what you eat, and I'll tell you what you know. The implications boggle the mouth as well as the mind. For can one really retain habits of claustrophobia and paranoia while burning with the endorphin-induced bliss of Mexican chili peppers? Can one sustain anger and resentment while absorbed in a French pâté de foie gras? And nibbling Japanese bites of sushi crowned with pickled ginger, lent fire by wasabi horseradish and brought to a threshold of ecstasy by sips of hot saki, makes the balance of trade seem a trifle to be resolved over the soothing comforts of green tea ice cream. An exaggeration, you say? Of course, but one devoutly to be wished and a fitting fillip to the desperate measures taken by desperate nations. Bring these nations together at table, soften their unbending postures through the most ancient and most available of seductions, the ravished palate, and somehow rigidities fall away in the prima materia of a perfect roast chicken.

In the midst of all this, fantasy and whimsy will have their day. *The Atlantic Monthly* recently sponsored a contest called "Combination Platters," the results of which made the rounds on the Internet. Readers were challenged to invent a new kind of international food or to name a restaurant that specializes in an interesting fusion cuisine. For instance, readers suggested, Middle Eastern and French pastry might be served at a café called The Fertile Croissant; a Franco-German romantic hideaway could feature Stollen Quiches; a Jamaican-Jewish salmon dish would naturally be called Dread Lox; while that most popular of all Japanese-German wines might be named Sumo Riesling. What a world! Wok and roll, indeed.

I have always been a fusion cook. Early experiments in creating dishes that my Southern-fried everything, garlic-hating father and my pasta-bred, olive oil–loving Sicilian mother would both appreciate got me started at an early age. Through my travels, I have developed a palate that savors the rich and varied flavors of the world table. Naturally, I try to bring together ingredients and techniques widely separated by geography and culture in the dishes I prepare. In the process of inventing my own recipes, I have discovered that if you mix cuisines you change the templates of consciousness. This is not exactly to say that a limited cuisine limits the mind and its matters, but rather that greater variety in foods and their fixing quickens the centers of awareness to potentials and perspectives not generally known to those whose diet is limited to regional equivalents of bread, meat, and potatoes.

And now, with the marriage, or at least the romance, between far-flung ethnic foods and their preparations, the possibilities for transformation are even more savory. World cooking is awakening us not only to other cultures but to the symbolic as well as the physical properties of food. Global cuisine quickens our appreciation of natural medicinal herbs and spices, perhaps the lost key to longevity, long known to indigenous peoples, engendering a renaissance of the ancient healing heritage of the human race. Each culture, we are coming to discover, has its own intrinsic sense of the life-giving energy of food. The palate of new cuisines spread across the world table is a hologram bearing the secrets of an ancient body-centered wisdom, triggering a new understanding of the universality and depth of who we are.

Go into most fine restaurant kitchens in North America and you find Japanese chefs trained in France, American chefs who have studied in Shanghai, and a rainbow coalition of sous chefs who have mastery of many cuisines. Practitioners of a high art form, the best of the new fusion chefs are blending the notes of individual cuisines into a symphony of culinary sound never heard before. The complex music of the new cuisine is so unexpected that eaters are enticed to leave behind the sensory expectations of the familiar and take off for new territory. A great fusion meal demands that we meet the world's flavors from a new place in consciousness. When we eat familiar foods, we know what to expect; our response is comfortable and well established. Take, for instance, the homey warmth and slightly sweet and silky perfection of a steaming bowl of French onion soup. Then mix in cilantro and lemon grass and even an unusual root vegetable, like purple potatoes. Suddenly we have to be cognizant of what we are eating. Jolted out of our sleepy state, we wake up to a variety of tastes and textures, a new global geography of place—the French Alps conjoined with the fields of Thailand and a Peruvian garden. Putting the basic building blocks of taste together in new ways is a reveille to the senses, an innovative and engaging wake-up call.

If prepared and eaten consciously, fusion food is transcendent and transforming. It engages the child in us, awakening belief in new things. It grants us a return to innocence. Of course, the sublime can also become the ridiculous. When fusion food is created randomly, as happens in too many restaurants, it is like a monkey hitting the keys of a piano. In one pretentious New York dining spot I was served a southwestern chicken taco with pineapple and sundried tomatoes muddled in a mango sour cream sauce. To add insult to injury, the waiter advised an accompanying chocolate martini. But as practiced by the masters, a fine fusion dish is like a fugue, not only on the gustatory level but on the spiritual as well. Eating, then, is like ascending the steps to a sacred temple, each taste taking us higher and higher, up to another level of consciousness.

Here, for example, is a consciousness-changing dish. It carries elements of many diverse corners of the earth. Let us call it The Salmon that Swam around the World. On a large white plate is a lightly grilled salmon fillet. The silky orange fish is topped with wild mushrooms, a mix of porcini and shiitake, which have been quickly sauteed in clarified butter. Drizzled over it are lines of a light green sauce compounded of coconut milk and cilantro and topped with grated fresh ginger. On the side of the plate is a limpid pool of golden-yellow curried yogurt sauce with mango chutney. Framing the presentation is a ring of tiny haricots vertes, exquisite French green beans, and the whole is nestled on a bed of ancient grains, amaranth or quinoa, cooked in chicken stock and butter.

This dish is structural as well global. It brings together salmon from the Irish seas, a fish of mythological importance as the bearer of wisdom. The wild mushroom blend fuses Italy and China, while reminding us of Hecate and her esoteric gnosis of the crossroads, food grown in darkness, the gathering of ancient awe and wonder. Coconut milk brings perseverance, what with the energy and focus required to break open the coconut and gather its tropical sweetness. Cilantro, a clean, clear, problem-solving herb, wakes up our mental alertness. Ginger is a powerfully healing food as well, bracing and astringent. Then India enters with its fiery notes, marrying the curry flavors of many pungent seeds with the soothing sweetness of mango bathed in yogurt. This combination not only wakes up the digestion but grants us sufficient fire in the belly for insight and the stomach for risk and adventure. The surround of bright green haricots vertes honors spring and renewal. At the base, as should be so, is the most ancient of grains, carrying earth energies, the gathered wisdom of the grasses of our planet, grown for twelve thousand years.

Such cooking gives us the wisdom of the future with the knowledge of the past, an incredible time consciousness that, like the salmon who returns to where it was spawned, helps us find our way home to the land of our future self. At its best, fusion cuisine is a roadmap to the cellular memory of our species—a virtual tarot of food.

World Music

As transformative as fusion food in today's world is fusion music. To listen to music over the last several decades is to be a Mendelian witness to a proliferation of genetic changes worthy of the old monk's generations of hybrid peas. Music by its very nature mixes and mutates according to the sounds of the shifting world. The globe is garrulous, but even more, it sings with the music of all its peoples.

Say you are a transgalactic musicologist and have occasion every hundred years or so to fly over the Earth, picking up its sounds. A hundred years ago you would have heard each region echo with a unique musical style: the percussive multirhythms of Africa, the chansons of France, the plaintive lilt of Irish laments, the sitar meditations of India. You are stunned at what rises from the Earth at the beginning of the third modern millennium. Hovering over North America, you marvel as Fred Ho and the Afro Asian Music Ensemble blend world music and progressive jazz into a soulful fusion of Duke Ellington, Chinese opera, Count Basie, John Coltrane, Asian folk songs, Charles Mingus, and West African rhythms, leavened with the radical politics of Malcolm X and a few scatterings of Mao Zedong. As you skim across the continent, you recall that the nation's capital city a hundred years ago was a bastion of banjo or Bach. Now you observe that the Kennedy Center is celebrating a musical synergy of the experiences of Africa and Spain as they met in the Caribbean's outer islands. The lines between classical symphonic, jazz, flamenco, and be-bop blur in a rhapsodic and buoyant celebration of the merging of West African Yoruba deities and Catholic saints. "Far out," you say in the language of your planet and head your ship toward Asia. Ah, there is something recognizable, but it's in the wrong place. Puccini's opera *Turandot* performed in Beijing's Forbidden City, staged by a Chinese director, with a Florentine orchestra and chorus, a maestro from India, and gorgeous silk costumes inspired by Chinese opera. Coasting down to Central Australia, you listen with wonder as an Aboriginal ensemble performs its own merger of contemporary and really ancient "rock"—stones beating, sticks clicking, and the hollowed out branch of a tree emitting a weird and wonderful drone amid the electronic pulsing of synthesizers, gathering the heart and politics of a people living between two worlds. "Back in another hundred years," you exclaim as you hit the hyperdrive button. "The universe only knows what they will think of next."

Next is now, world music as the single most potent force in the culture of fusion. How this is happening is a piece of radical sociology. Immigrant populations establish their identity and psychic health in cities throughout the planet by creating music that upholds the soul of where they came from while adopting some of the forms and formulas of the mainstream. What results is traditional music redeployed for untraditional purposes—authentic ethnic sounds metamorphosed into global pop.

Take Sheila Chandra, for instance. Born in England of South Asian parents, she refuses, as she says, to be an "Indian living museum piece here in England." Rather she thinks of herself as a "world citizen," and indeed her music combines ancient Islamic and Irish vocal styles with traditional music, as on her breathtaking album *Weaving My Ancestors' Voices.* Surprisingly, it works very well, an organic fusion of two cultures that rarely meet; when they meet here in song, they become gateways to the meeting of cultures in physical space. Thus are new geographies opened and domains of previously divided and distinguished worlds mutually enchanted through the magic of Chandra's gift for universalizing through the voice. In *The Zen Kiss* she offers "Speaking in Tongues," a piece in which she adds her own experimental aesthetic to the punctuating tongue-twisting syllables voiced by male tabla drummers playing in the Hindustani tradition. On the same CD she pursues her passion for empowering women with a piece of plainchant drawn from the music of that avant-garde medieval mistress of cultural forms Hildegard of Bingen. Chandra adds her own unique multicultural effects to Hildegard's music, principally a drone sound that fashions an undercurrent of blood music that joins musical soulmates separate in time but not in spirit.

The music of the diaspora is not just the aesthetics of crossover but the powerful agent by which the local disrupts the global, tradition deconstructed in the service of a new world commons. World music–makers are like shamans, carrying us on magic carpets of riffs and melodies through states of consciousness that spin us out of time to lands not yet invented but glimpsed on the aural horizon.

Of course the original fusion music of our era is the music of the African diaspora, which gave birth to such disparate musical styles as rock, jazz, blues, reggae, salsa, samba, bossa nova, juju, highlife, and mambo. As a result of the Atlantic slave trade, ancient African principles of song and dance crossed the seas from the Old World to the New. Here they took on new momentum, intermingling with each other and with New World or European styles of singing and dance. The music that resulted from this union is a coding of creativity and imagination, of religious feeling and interwovenness of spirit and matter that is as exhilarating and evolutionary as it is fearful and awesome. In African metaphysics, there is no Great Divide; the human world and the spirit world intersect, feeding each other, and hotting the world up in the meeting. The metaphysical goal of African-inspired music is the experience and celebration of the ecstasy of being the vehicle through which the human world and the spirit world interconnect.

The drum is the key to this spiritual encounter; the dance is its instrument. In Africa, to meditate is to dance. In West Africa I have seen men and women of all ages, from year-old babies to very old people, praying by doing a kind of dance movement in which the thoracic cavity is continually flexed and opened up to receive the god or goddess to the beat of the drums. The drumming and dancing allows the body to become a meeting place for spirit and a vehicle for immersion into ecstasy. In Mother Africa, the holy temple is not some architectural structure; it is the body itself that is pulsed into theophany. Ego is obliterated, as is any sense of self or local concerns or conditions. One has gone south in one's psyche. The African gods are thought to be dancers themselves, that is, frequency waves or rhythms that are closer to the Great Rhythms and Great Patterns than we are. To dance, then, is to enter into communion with the larger dance that is the universe. And because the universe dances, "he who does not dance does not know what happens."

Michael Ventura has written passionately in *Shadow Dancing in the USA* about why African-inspired music and dance has had such a potent impact on Western culture:

> Jazz and rock 'n' roll would evolve from Voodoo, carrying within them the metaphysical antidote that would aid many a twentieth-century Westerner from both the ravages of the mind-body split codified by Christianity and the onslaught of technology. (Los Angeles: Jeremy P. Tarcher, p. 120)

In the diaspora, the Voodoo metaphysic moved from explicit practice to implicit wordless power in the music that was born of it—jazz, blues, rhythm and blues, rock-and-roll, and gospel singing. Some of the practices of the path would be preserved in the lyrics, especially the recurrent theme of the archetypal long snake called Damballah. The music with its sinuous long snake, DNA, spiral galaxy, kundalini, vortex pattern forms seems to be stating in rhythmic terms the themes of our new contemporary reality.

Come the late 1940s, and southern rhythm and blues musicians like Carl Perkins, Jerry Lee Lewis, and Elvis Presley were listening hard to largely black early versions of rock-and-roll, as interpreted by Big Joe Turner, Muddy Waters, Willie Dixon, Little Walter, and Sonny Boy Williamson. In these African-inspired sounds, these white musicians heard the wild haunting music, the long snake moaning, and by the 1950s, they too began to moan. Here was Elvis, a white man, singing, dancing, and playing black. With Elvis, you not only saw the moves he made with his body, you heard them. He leaped and twisted and shimmied and shook into directions beyond the normal three-dimensional ones. His groin propelled itself into the fifth dimension, to make girls moan and boys imitate. When the student is ready, the teacher will appear, the old saying goes, and Elvis appeared on the scene in spite of himself to fill the needs of many disciples. Although the establishment could ignore black musicians, it couldn't ignore Elvis. And after Elvis, Little Richard, Fats Domino, Chuck Berry. Though American culture had been, up to then, trying to damp down the ecstatic forms of cultures that came from across the sea, it could do so no longer. Albert Metreaux has said of Haitian voodoo dancing:

> Spurred by the god within him, the devotee throws himself into a series of brilliant improvisations and shows a suppleness, a grace and imagination which often did not seem possible. The audience is not taken in, it is to the loa [god] and not to the loa's servant that the imagination goes out. (Quoted in Ventura, p. 154)

Sometimes in America, as Ventura reminds us, we worship the servant rather than the loa, with the result that the servant gets blown out by our adoration and goes on drugs or drink or worse to try to capture some of the archetypal sense of self that is being so mightily projected onto him. This possession by the loa or god became the archetypal form of American performance in rock-and-roll. Elvis Presley, Little Richard, Jerry Lee Lewis, James Brown, Janis Joplin, Tina Turner, Jim Morrison, and, later, Bruce Springsteen "let themselves be possessed not by any god they could name but by the spirit they felt in the music" (p. 154). And their possession state moved out in resonance waves and possessed their hearers powerfully, even through records and cassettes. Often these recordings were epiphanies without metaphysics, trance dances without morals. However, the body got the message, and the Western body image was forever changed, made more aware, feistier, juicer, sexier than it ever had been. Rock-and-roll has provided for a critical antidote for the Western body-mind split, and the energy of the spirit that fills the performer now fills vast audiences with an effect that is as great in its consequences as any historical event in the last three hundred years. With its influence on youth, the groundwork was laid for potent changes: the movement against war and nuclear power, the civil rights movement, feminism, ecology, the whole new age, and yes, the appreciation and study of Eastern philosophies in the East-West axis and the shamanic knowings of the North-South axis. And so we see that music brings us, in the jumping of our cells, from what is past to what is trying to become the future. It expresses and fulfills our need to hear the spirit in the dark. In its musical legacy, mysterious Africa has had consequences of greatest light.

Globe-Wandering Writers

The shift in consciousness mediated by fusion music is paralleled by provocative trends in recent literature and theatre. With English now the official language of sixty-three countries of the world (while in the Los Angeles school system, eighty-two different languages are spoken), a new kind of voice with a different ancestry and accents cooked in disparate climes is beginning to shape the world of letters, particularly English literature. As with music, this shift is mediated by the fluid movement of writers around the globe, such that Michael Ondaatje, a native of Sri Lanka of mixed heritage, can set his prize-winning novel *The English Patient* in a Tuscan hill town and people it with characters from India and North Africa as well as from Europe. Or take Richard Rodriguez, a son of Mexican parents, who feels himself to be both an Indian and an Irish Catholic and who lives and writes in America with the intrepid freedom of an outsider. In his recent work *Days of Obligation,* a Pakistani converses with a Mexican in a Chinese restaurant, and Native American moccasins are manufactured in Taiwan.

Thus contemporary literature has become a continuous movement to different realities as well as to different shores, as essayist Pico Iyer writes: "children of many cultures—born 'multiculturalists' who do not need to use the word—they show us how we can create our own traditions, free of labels, in a universe of our own devising" (*Tropical Classical,* New York: Vintage, 1997, p. 124). Iyer, an elegant, globe-wandering writer, comes by this hard-won knowledge firsthand. He is the son of Indian parents brought up in Oxford; his prose yokes his classical education in a formal English boarding school and his family's move to freewheeling California in the 1960s with the tropical palms still waving in his family's memory. Iyer terms the worldscape of the best of fusion writing "tropical classical" and points to poet Derek Walcott, novelist Michael Ondaatje, and essayist Richard Rodriguez as "three of its masters." Their particular genius, Iyer says, is

> the ability to season high classical forms with a lyrical beauty drawn from the streets and beaches of their homes. To learn from the tradition of Homer and Herodotus and Augustine, respectively, and yet to enliven and elevate those dusty forms with the rhythms of Saint Lucia, the colors of Sri Lanka, the love songs of the Latin South. (*Tropical Classical,* pp. 123–4)

Rudyard Kipling once wrote, "The wildest dreams of Kew are the facts of Kathmandu"—to which one can now add the present inverse truth that the wildest dreams of Kathmandu are the facts of Kew. I have visited Kew as well as Kathmandu and can testify to their current crosspollination. The old hippies of Kew foment revolution while sitting in a psychedelicatessen of drugs that promise a palette of potential nirvanas, while certain denizens of Kathmandu seek enlightenment visualized as a Mercedes and a Rolex. This exchange of dream and reality is producing writers who are at home in many cultures and who have an eye and ear for its polyglot world. Yet these expressions are ultimately a celebration of creative freedom. As Bombay-born novelist Salman Rushdie writes in *Imaginary Homelands,* "We can't simply use the language the way the British did; it needs remaking for our own purposes. . . . To conquer English may be to complete the process of making ourselves free" (New York: Penguin, 1992, p. 17).

Rushdie and the other chroniclers of the fusion landscape often wrestle with the puzzle of who they are and revel in the freedom and consciousness gained; they are evolutionary experiments in search of an author who is, of course, themselves. Living in the in-between, with multiple identities, they awaken in the reader dormant arenas of consciousness, subtle mine fields that can be tripped by the exotic phrase or juxtaposition of unlikely characters. Rushdie's characters reflect a polyphony of cultures; his images are bountiful, an exotic outpouring from the cornucopia of his fusion imagination. He says of his own work: "One of the images I find most satisfying in *Midnight's Children* is the image of leaking, of people leaking into each other, like flavors when you cook" ("A Conversation with Salman Rushdie," interview by Una Chadhuri, *Imaginative Maps,* New York: Turnstile Press, 1990).

Rushdie's own experience often leaks into his novels. When the Ayatollah Khomeini issued a *fatwa,* or death sentence, on him for "polluting the name of God" in his *Satanic Verses,* Rushdie lived for years under constant threat of assassination, moving nightly from one place to another. His latest novel, *The Ground Beneath Her Feet,* could be an allegory of his recent life, for the ground shifts repeatedly beneath his characters, often literally. The novel's heroine, Vina Aspara, the daughter of a Greek American woman and an Indian father, is the lead singer in the most popular rock group in the world. As the novel opens, she is swallowed whole by the terrible earthquake in Mexico City, descending to the underworld like a modern-day Eurydice. Her Orpheus is composer-performer Ormus Cana, the Bombay-born founder of the rock group. Their epic romance of love found, and lost, and found again, stretches through flashbacks from cosmopolitan Bombay, to London in the vibrant 1960s, to New York, to the world beyond death. Vina apparently reincarnates and confirms Ormus's messianic knowing that "there is a world other than ours, and it's bursting through our continuum's flimsy defenses" (New York: Henry Holt, 1999, p. 347).

Rushdie's evocation of the permeable borders between intersecting worlds approaches the reality known to my Aboriginal friends in Australia. During the time I stayed with them in their camp in Arnhem Land, they told me that our world is nested in another world—the Dreamtime, *Ngarungani,* the depth world flowing through this one. It is always and never; a time that never was and is always happening. It is the force that spiritually maintains the world and has to be constantly recreated through ritual and prayer if the world, the clan, and the self are to be renewed. The Dreamtime includes a mythological past when creative spirits moved over the land, shaping it, naming it, calling it into being, and generally getting it ready for human habitation.

In evoking their own particular form of the Dreamtime, fusion writers like Rushdie call on ancient myths of the time before time—love and death, creation and destruction, the odysseys of the soul. Like the creative spirits of the Dreamtime, they prepare the ground for their readers to break through into a new world, bridging the now with the surreal and fusionary realities of the soon-to-be. Amid the shifting sands of the fusionary landscape, the dim outlines of a new world are being brought into being, one in which nationality, race, gender, history, myth, and political realities are rendered immaterial, and the world is transformed into a stage set upon which the tangled plot of the future is unfolding.

Perhaps my sense of history as an unfolding drama is why I am such an inveterate playgoer. Wherever I am in the world, I try to attend as many performances as I can. After lectures or seminars, I am asked by my hosts, "Jean, can we take you to dinner?" "No, to the theatre, please," I invariably reply. As a result I have seen madcap fusionary productions of *The Tempest* in Caracas and of *Hamlet* in Stockholm (staged in eerie but bawdy futuristic style by Ingmar Bergman). Closer to home, at the Next Wave Festival at Brooklyn's Academy of Music, I witnessed perhaps the most wonderful fusion drama of our time, the strange Hermetic magic of Peter Brook's nine-hour production of the *Mahabharata*.

Translating the *Mahabharata* into a contemporary stage piece is already an exercise in fusion alchemy. Dating from 400 B.C., this voluminous epic collection of Hindu history, mythology, and thought tells of the complex struggle between two opposing sets of cousins in an ancient Hindu dynasty. Brook staged the work as a veritable tour of Oriental and European minimalist stagecraft. The battles of multitudes were conveyed by acrobatic displays of Eastern martial arts; brightly patterned carpets and swirls of red and gold fabric carried the audience into the realm of the gods; and puppet sequences, masks, and mime drawn from Japanese and Balinese theatre helped actors morph into god and animal hybrids. Originally written in French, the play was translated into English for the New York production; the large international cast included actors of all races and ethnicities, and for many of them English was a second language. Allusions to the Old Testament, to the Oedipus cycle of Greek drama, and to Shakespeare—which is Brook's perennial fascination—linked the ancient Hindu story to the classics of Western literature.

In the 1960s, after more than forty successful productions, Brook abandoned the most innovative and brilliant theatre career of his generation to explore the primal roots of drama in the myth and ritual of many cultures. Leaving conventional theatre behind, he created an institute in Paris devoted to research into the ancient and arcane. In so doing, he became a modern avatar of Hermes, a guide to the mysteries of the cosmic theatre in which we all are stars. Brook believes that audiences should be drawn into remembering the better parts of themselves. Thus one is a spectator in name only. As his plays progress through the night, one feels a part of the mystery tradition of ancient times—the rites of the ancient Mysteries, which provided powerful initiatory journeys of anguish, grief, loss, redemption, knowledge, and ecstatic union with the god or goddess. During the journey one "died" in some sense to one's old self and was reborn to a higher self or even to identity with goddess or god. One also received special knowledge and training as part of one's initiation into a deeper life and its meaning. A fragment of a work by the Greek poet Pindar says of the initiates upon their return: "Happy are they who, having beheld these things, descend beneath the Earth. They know life's end but also a new beginning from the gods."

To witness the great mythic dramas as staged by Peter Brook is to descend and ascend in the same evening, an initiate in a once and future Mystery, igniting the power of life beyond death. In calling Peter Brook a modern Hermes, I am putting him into the company of those messengers who bring the news of Jump Time. In Greek mythology, Hermes is both the messenger of the gods and the guide to the underworld. He is therefore the one who brings together radically different cultures—gods and humans. He is also the god of *kairos,* the loaded time when change and transformation can happen. Hermes brings the windfall, the unexpected but fortunate happening at the right moment; in the Odyssey, for instance, Hermes shows up to give Odysseus the plant *moly,* which will prevent his being turned into a pig by Circe, and Hermes brings the message to Calypso to let Odysseus off her island where he has been stuck for seven long years. Hermes can also be mischievous, a trickster. Later, he assumed the role of soul guide, or psychopomp, to the Mysteries and to the depths. He is therefore a suitable catalyst for Jump Times.

The Cosmopolitan City of Alexandria

There's a potent connection between Hermes and the mysteries of Jump Time. In the second and third centuries of the Common Era, Hermes came to be associated with Thoth, the Egyptian god of magic and mystery, and, as Thrice-Great Hermes, Hermes Trismegistus, he became the central figure around whom a fusion developed of Platonic Greek, Egyptian, Babylonian, and Hebraic writings on natural and occult science. The study and interpretation of these so-called Hermetic texts was an important strand of philosophical and spiritual work for the intelligentsia of the famed city of Alexandria, Egypt, who influenced and inspired a crosscultural intellectual ferment in the centuries following the birth of Christ.

Alexandria—the very name has potent charm. Once in the Middle East, I met an old man who told me, "In my grandfather's grandfather's grandfather's time, the strong blond man from the West came here with many men and horses. His name was Iskander. Some of his men stayed and married our grandmothers, which is why I have blue eyes." As I listened to the mythically embellished tales of "Iskander," it became clear that the old man was describing his family's distant memory of Alexander the Great. Follow the path of Alexander's conquests in the fourth century B.C.E. from Macedonia south and west to Greece and eastward through Persia, Libya, Egypt, and all the way to northern India, and you find many little towns named Iskander or some variation. All attest to the Macedonian warrior's policy of colonization and of marrying his soldiers to the local women. The cultural fusions that came of these joinings can still be seen. In India, Alexander's influence is evident in the saris worn by the women, thought to be an improvisation based on Greek and Macedonian women's costume of the time.

But it was in the vast city of Alexandria, which Alexander himself helped to design and lay out in 332 B.C., where the philosophical, scientific, and spiritual traditions of East and West fused into a cultural synthesis unmatched until the Jump Time of today. If you walked the streets of old Alexandria you saw Greek philosophers, Egyptian merchants, Mesopotamian gardeners, Roman soldiers, Celtic jewelers, Hebrew scholars, Persian perfumiers, African weavers, Indian physicians, and even a few Buddhist missionaries—a veritable hothouse of peoples and ideas. This was fed by the gradual collapse of political boundaries within the Roman Empire, which gave rise to a relative freedom of movement. Egyptians could take up residence in Greece; Syrians in Rome; inhabitants of Asia Minor in Gaul; Africans in Spain; and everybody from everywhere came to Alexandria. One was no longer a *polites,* or citizen of a local city, or *polis*; one had now become *kosmo-polites,* a citizen of the world, a cosmopolitan, free to pursue the spiritual and psychological realities of other cultures. Looking at one thread of this cultural exchange we can see how the spirituality of the time, once tied to a state or regional cult, bloomed with many lavish blossoms. Archetypes and universally resonant symbologies developed. Thus religions that had grown out of the old civil state religions became, in the cosmopolitan reality, more varied and psychologically evocative than they had been in their provincial form.

As the supreme *kosmopolis* of the time as well as the greatest and busiest international seaport, Alexandria was a cauldron of cultural exchange. Its famed library held more than half a million papyrus scrolls, the largest repository of information in the ancient world, sufficient to support and sustain a heady brew of cosmological, literary, mythic, and metaphysical thought. In the six centuries of the library's existence, from three centuries before Christ to three centuries after, the greatest scholars of the day debated in its meeting rooms, speculated in its astronomical observatories, and strolled in its zoological and botanical gardens. Among them were Apollonius the poet, Euclid the mathematician, Ptolemy the astronomer, Plotinus the Neoplatonist, Hypatia the philosopher, and even the Christian theologians Clement and Origen—lives that pushed the edges of the known because they were nurtured in the rich and varied stimulus of world culture as it was known at that time, the geography of the juicing mind.

Alexandria was thus a teacher's as well as a student's paradise—the city as multiversity, where virtually anyone could have access to the best thinking of every known tradition, and ideas unfolded and were blown from mind to mind like the gentle winds that filled the sails of the ships bearing news from distant seas. Before the Internet, before the global village, Alexandria was, for a time, the prototype of a universal ethos, a place where a tolerant pluralism flourished and the widest and deepest currents of thought and belief could be explored. There, any dogmatism was laughable, for were not the varieties of this myriad world emanations from the One?

It did not last of course. When Christianity became the official religion of the Roman Empire, some Church Fathers ordained that knowledge was dangerous and that only the Christian doctrine held the truth and the way to salvation. Classical literature and learning were banned, other religious traditions were prohibited, and non-Christians had their property and sometimes their lives confiscated. The most infamous event was the murder of the female Neoplatonic philosopher and teacher Hypatia. Her beauty, it was said, was matched by her brilliance in mathematics and astronomy and by her virtuosity in logic and debate. So luminous was her mind and manner of teaching that she attracted Christian students as well as seekers of knowledge who came from all parts of the Roman Empire to study with her. In the face of growing envy and opposition from the Christian patriarch Cyril, she courageously pursued her scholarly life until she was attacked and torn to pieces by a frenzied Christian mob. Such toxic fundamentalism shattered the greatest teaching-learning community of cultural fusion the world had ever known. One wonders whether the Dark Ages would have been so dim and life so diminished if the spirit of Alexandria had been allowed to flourish. What would the world be like if Christians had permitted the peaceful coexistence of the "new testament" and the ancient wisdom of the pagan world, just as Hermetic philosophy and the worship of Isis, or the study of mathematics and the pursuit of medicine, had so brilliantly dwelled side by side in Alexandria's fusionary society?

And yet Alexandria at her best—passionate, protean, and a

dynamic field of fertile fusions—remains a Platonic archetype in the ether of social forms, forever modeling the way in which a multicultural society can fan the mind and the spirit into creative riches, profound investigations. Today we meet this meta-Alexandria looming as a benign presence over the Internetted, intercultural world. The library of all conceivable knowledge is at hand, not just through interlibrary loan but through modem-mediated access. Give me a half a day on my computer, and I can get you hundreds, even thousands of pages of material to peruse on any subject, from Airedales to astrophysics. And so can you. The contemporary cosmopolis is the global village with its meeting and melding of cultures and ways of being. It is a virtual community, in which everyone has access to everyone else and to something more—the universal spirit that seems to be luring us into a new kind of unity, prismatic in its many colors and flavors, deepening our individuality while we move toward a unitive vision of our wholeness in the light of the One.

Fusion's Backlash

The story of Alexandria gives us warnings in the manner of its demise and, perhaps, signals that suggest ways we might prevent similar collapse and regression. The down side of our own time's cultural fusion is that the sun of all this rapid becoming casts long shadows, fear and mistrust before a level of change unexpected and inexplicable. As in ancient Alexandria, the arbiters of entrenchment and insularity are many, their arguments convincing to a surprising number. Newspapers and newscasts daily give us evidence of backlash that too often crosses into mania. Nazi skinheads, white supremacists, murderers of gay people, batterers of women, bombers of government buildings, militiamen, fundamentalists of every stripe and order are displayed at every turn. When the mind-ways of decades, centuries, and even millennia are suddenly seen as being compromised by the incursion of the new, then the old brain rears up its ancient head and says, "Nothing doing." Initiatives in both the United States and Europe for white people to "take back their country" parallel exploding ethnic fundamentalism in other parts of the world. India has its own version in the hyperconservative Hindus who are rabidly anti-Muslim and anti-Western. They protested vehemently when a recent episode of the international television hit *Xena* had the warrior princess confer with Krishna and herself turn into a multiarmed Kali. Demonstrations and threats by over a hundred Hindu organizations over what they considered a desecration of their theology and sacred icons forced the producers to withdraw the episode from distribution.

In a less dangerous way, many professional pundits dismiss the phenomenon of cultural fusion as merely the current manifestation of conspicuous Western consumerism. They disparage attempts to understand other cultures and their forms as vapid visits to the Disneyland of sightseeing—cultural fusion as a T-shirt logo—"Hard Rock Cafe Marrakech." Moreover, they point out, corporate sellers massage the marketplace with commercials that employ an exotic spray of multicultural sound bites, a "technopoly" in which all forms of cultural life are processed by crass technique and technology—Brazilian soccer stars selling Nike shoes made in Vietnam to American ghetto kids or Coca-Cola selling merchandise via remixed Muslim hymns.

America is the ever-spawning womb of pop culture, sending its hybrid kids out like mayflies to give the latest buzz on life in the Promised Land. When you have watched *L.A. Law* in Kyoto and *Dynasty* in a small village in South India; when you have read how Laplanders put off their annual reindeer migration by a week so that they can find out who shot J.R., you know that America and its popular media has a stranglehold on the hearts and the pocketbooks of the world. Gore Vidal tells us that the capital of the world is Hollywood, and why not? For America is the dream factory, the center of all image-making. American pop culture is the matrix of all possibilities, the archetypal warehouse for music, sport, fashion, fast food, finance, gossip, films. Politicians come and go, their brief gray eminences receding in the perpetual sunlight of Madonna and the Michaels, Jackson and Jordan.

And so we have "America: The Myth," which has become as unreal as "Reality: The Movie"—a liminal, not-yet, never-really-was, but maybe-soon-to-be place that serves as the Dreamtime to the world at large. Fortunately or unfortunately, as the case may be, America occupies the largest continent on the planet of the world's soul. But curious things are happening in this planet's ecology. Whereas cultures and people are becoming hyphenated in their psyche's address, for much of the world the other end of the hyphen is often "American." As Thomas Friedman writes, American globalization has become the world's uninvited guest.

> You try to shut the door and it comes in through the window. You try shut the window and it comes in through the cable. You cut the cable, and it comes in on the Internet over the phone line. When you cut the phone line, it comes in over the satellite. When you throw away the cell phone, it's out there on the billboard. When you take down the billboard, it comes in through the workplace and the factory floor. And

> it's not only in the room with you, this Americanizaton-globalization. You eat it. It gets inside you. (*The Lexus and the Olive Tree,* New York: Farrar, Straus & Giroux, 1999, p. 319)

This trend creates its own polarities, and many cultures, after decades of buying into the ambiance of the world's only superpower, are searching out and asserting their native traditions, reclaiming their authentic identity and its inherent richness. Indigenous peoples around the world are bringing back their language, dance, music, oral traditions, healing modalities, spirituality—all those things that distinguish them from other cultures, and especially from America. The exquisite irony of all this is that Americans have served as traveling Medicis, providing a market for the resurgence of world cultures by sitting down in Alice Springs for didgeridoo lessons from Aboriginal elders, commissioning intricate Balinese carvings in Ubud, and holding still for Ayurvedic healing treatments in Trivandrum, South India.

Part of the mystery of America is that it is the paradigm of nations, beginning as it did as the visionary land of opportunity that was to fulfill humankind's millennia-old dreams of the Golden Age returned in a new land beyond the Western waters. But now something else is going on. The many cultures that have come to America have charged the soil of our national psyche with the potency of thousands of years of imaginings and expectations. Now as our hybrid and hyphenated American culture exports itself to the world, it carries with it some of this psychodynamic charge, a dream that by its very success activates a yearning for self-expression in cultures around the world. Maybe in a contrary and surprising way, America is serving Gaia's grand ideal: as the Earth is hourly losing species, perhaps she needs to deepen and strengthen those human cultural species that remain, so that a vibrant and various ecology of peoples may propagate and green the coming global civilization.

Where will it go? How will it end? Perhaps a clue is given in the marvelous words of Edna St. Vincent Millay, from the poem "Huntsman, What Quarry?"

> Upon this gifted age, in its dark hour,
> rains from the sky a meteoric shower
> of facts . . . they lie unquestioned, uncombined.
> Wisdom enough to leech us of our ill
> is daily spun; but there exists no loom to
> weave it into fabric.

Those words were written in 1923, but now, in a new millennium, we have weavers—the artisans of many cultures, who together are building that loom. What threads they bring, what songs they sing, what tastes they savor are the strands of kairos, the potent moment when the fabric of Jump Time is beginning to take form.

CHAPTER 7

PSYCHENAUTS IN CYBERSPACE

As my hands fly over the keyboard, the sunlight streaming through the window washes out the words on the computer screen but lingers for a minute on my hands. A memory rises of other sunlit hands. They belonged to a man I used to call Mr. Tayer, whom years later I discovered to be the great Jesuit paleontologist and mystic Pierre Teilhard de Chardin. He lived several blocks from me, and when I was in my mid-teens and he in his seventies, we took walks together occasionally with my fox terrier Champ in New York's Central Park. I have told the story of these encounters many times and written about them in several of my books. But as I have come to realize, I can never tell this story often enough, for it was, in a mysterious way, a Jump Time for me, one that propelled me into the life I now lead. I mention Teilhard here because, in the last few years, he has become the patron saint of a bevy of Internet theoreticians. Writing in *Wired* magazine (June 1995), Jennifer Cobb Kreisberg observed, "Teilhard saw the Net coming more than half a century before it arrived."

I am remembering a time when Mr. Tayer lifted his hands with their long fingers up to the sun and spoke in his interesting French accent about what he called the noosphere. For a moment, the light seemed to make his hands translucent. With his hands sweeping across the sky, he told me that the Earth had grown itself a new skin that encircled the planet. This new skin—the noosphere—was a vast thinking membrane and would one day be "the living unity of a single tissue that would contain all of our thoughts, our dreams, and even our experiences." It would be Mind at Large, the weaving of the consciousness of the planet, fueled by human awareness and quickening the human evolutionary journey.

"When will this be?" I asked him, already anxious to find some Know-It-All place I could go to for help with my high school term papers.

He told me that it was already in place and had been since human beings became self-conscious. But in my lifetime, perhaps, this living membrane would grow in density and complexity, activating the human species to greater consciousness and responsiveness. His words became strange, grand, as luminous and illusive as the sun rays that seemed to provoke his reflections. He spoke of the immense and growing edifice of matter and ideas, of the phosphorescence of thought and the irresistible tide of intelligence and spirit that was bringing about a change of planetary magnitude. He spoke of "noogenesis," the evolution of a new layer of life that was above the biosphere of Earth's living systems.

"But what about the trees and the rocks and the animals?" I asked, worriedly looking at Champ. "Aren't they important anymore?"

He answered that they were as important as ever but were now incorporated into and crowned by the noosphere. They and we were all part of a cosmic evolutionary movement that was moving us toward metamorphosis into a whole new form. As this metamorphosis continued, we would leave our littleness behind. He spoke of the new electrical connections—radio, television, and these new room-sized computers—as the outer forms of this inner change.

Shortly after this talk, Mr. Tayer died. But his visionary words were etched in my memory. Years later I read about the same ideas in his book *The Phenomenon of Man*:

> A glow ripples outward from the first spark of conscious reflection. The point of ignition grows larger. The fires spreads in ever widening circles till finally the whole planet is covered in incandescence. Only one interpretation, only one name can be found worthy of this grand phenomenon. Much more coherent and just as extensive as any preceding layer, it is really a new layer, the "thinking layer," which since its germination . . . has spread over and above the world of plants and animals. In other words, outside and above the biosphere is the noosphere. (New York: Harper, 1959, p. 182)

A Netted Reality

What seemed arcane speculation in the 1950s about a complex field of mind and information circling the planet we now recognize to have been a prophetic vision of the globe-circling web of electronic information. Energized by human consciousness, the web grants us a netted reality by which everything and everyone is woven into a fabric of information, ideas, experiences, further dissolving the membrane that kept peoples and cultures separate and insular. The Net's high-tech communion is spaceless and not bound by the usual categories of time. One can hook in from anywhere—a café in Paris, a basement in Beijing, thirty-five thousand feet up flying over the Arctic, a boat in the Aegean, a cab in Kansas City. The Net makes one ubiquitous, allowing rapid travels through all of the known as well as many of the unknown worlds. It is the matrix within which cultures meet and propagate in new fusions and peoples exchange their social DNA at a remarkable rate. From this mating, a whole new species is being born.

Teilhard's speculations in the 1950s were prescient to be sure, but the idea that a web of energy links all that is can be traced back further still. Almost two millennia ago, the second-century Buddhist *Avatamska Sutra* contains a mystical vision of the ultimate energetic net (I have paraphrased the original):

> In the heaven of Indra there is said to be a network of pearls, so arranged that if you look at one you see all the others reflected in it, and if you move in to any part of it, you set off the sound of bells that ring through every part of the network, through every part of reality. In the same way, each person, each object in the world, is not merely itself, but involves every

> other person and object and, in fact, on one level *is* every other person and object. (Houston, *The Possible Human,* New York: Tarcher/Putnam, 2nd ed., p. 188)

The World Wide Web is a present-day incarnation of Indra's Net, a metascape of electrons, holographic in character, and, like its metaphysical parent, an interdependent matrix, the One and the many in an infinite dance. The Buddhist meditates to know himself to be interdependent with this cosmic dance; the Internet explorer potentially can discover the same truth via an electronic pathway that can be as radical to consciousness as the spiritual technologies of the meditator. In both nets there is the mirroring, even the interpenetration, of the whole by the member parts. This mirroring is reflected in another translation of the ancient sutra, which reads: "Thus each individual is at once the cause for the whole and is caused by the whole, and what is called into existence is a vast body made up of an infinity of individuals all sustaining each other and defining each other" (Francis Cook, *Hua-yen Buddhism,* University Park: Pennsylvania State University Press, 1977, p. 3).

Perceiving the interconnectedness of all beings during spiritual practice loosens the boundaries of one's reality to prepare for the realization of one's place in the cosmic dance. Traveling the energetic byways of the Internet leads to a similar stretching and loosening of the membranes that traditionally divide cultures, languages, sciences, religions, nations, races. Every time we log on, we participate in the creation of the global mind field. The planet is becoming self-conscious in all its parts through ourselves. Electronic circuitry has so wired the planet that within a few years, just about everything that the human race is doing or has ever thought about will be available at our fingertips, our hands at play on the keyboard enabling the human spirit to come at us in resonance waves.

Moreover, the Internet is remaking human culture. In the fourth millennium B.C., sophisticated cultures grew up along the great rivers—the Nile, the Tigris and Euphrates, the Yangtze, the Ganges. Today, a new and very complex culture is growing up along the Internet's great river of electronic information. The electronic revolution is returning us to a tribal world of instantaneous information, reorganizing social interactions and changing people's public and private behavior before they even know it is happening. The Net world is a second universe, a kingdom in our midst, with sights and sounds, landscapes and knowledgescapes, markets and amusements, romances and resources—many of which have never before been seen on Earth. It burgeons forth, this global Village of villages, gaining each hour more and more inhabitants, who live and move and have their being in a world that is nowhere yet everywhere.

We who inhabit the Internet's virtual outposts are fast evolving into new kinds of beings, our neural system and sensory receptors extended through space and time. Psychologies that have endured for millennia are passing away in a few hundred months. In response, the human psyche itself is expanding—even, I believe, being remade. This dance of metamorphosis is reciprocal; the Internet is changing us, even as we refine the technology that extends its reach. As our new body-mind evolves, everything else is changing, too: our body image, how we think and use language, our relationships and our sense of community, the ways we work and create, even our view of the nature of reality itself. The Internet promises to bring about as great an evolutionary change as occurred when people stopped depending on the meandering of the hunt and settled down to agriculture and civilization.

Several years ago I put out a call for help on seven Internet lists to which I subscribe. In my posting, I asked that people write to me about how the Internet impacts human potentials. I asked a number of specific questions. Is the Internet activating new or rarely used potentials? Does the Internet promote greater flexibility of thought or the ability to see patterns of connection more easily? What is its effect on the use of inner imageries? What is happening to visual, auditory, and kinesthetic thought? Are new frames of mind evolving? Does the amount of information on the Internet lead to overload? To one list focused on spirituality and technology, I asked: Does a forum like "Techspsirit" encourage archetypal and mythic imagery and thought? Is cyberspace engendering new mythic structures? What is the future of consciousness on the Internet? In response, I received over a hundred and forty replies. As you will see throughout this chapter, they provide firsthand testimony to the metamorphosis the Internet is bringing about in human capacities and culture.

The Internet and the Body

The first issue I addressed in my informal research was the impact of the Internet on the body. A number of respondents reported changes in their concept of their own bodies. Some felt disembodied while wandering the Net—ghosts in the well-known machine; while others—the majority, in fact—reported a sense of enhancement, even a greater embodiment. One man wrote me, "It's as if I have two bodies now, the one at the computer, and the one out there traveling the Net. Then, when I think about it, I call the Net body back, and I feel somehow thicker, denser, a doubled person. Does that make sense?" I wrote back suggesting that what he was feeling on the Net was an identification with his kinesthetic or imaginal body, which, as many dancers and athletes know, can feel as vivid and as "real" as the physical body.

Other respondents talked about becoming more aware that humans are prosthetic beings who extend their arms and legs through wheels and tools and machines, their nervous systems through electronic devices. Now, with computers and the Internet, this sense of the amplified body has taken a quantum jump. A college student who spends as much as forty hours a week on the Internet wrote:

> Sometimes, around 2 a.m., I feel myself to be a giant body, with nerve endings in Turkey and Quebec and Glasgow. I feed my netfriends with pieces of myself, information I've gathered, thoughts I've been thinking, and they feed me with pieces of themselves. My heart is in Italy, for I've fallen in love with a girl from Milan who talks to me regularly on the Net. The romance we have is deep and good, although physically, we've never met. My intellect is wired to many countries. I am majoring in economics, but find my college professors nowhere near as informative or useful as an editor of the British magazine *Economist* I've met on the Net, or a cyberfriend in France who has come up with a whole new way of looking at economic redistribution. I am doing my senior dissertation on his theory and its implementation. So you see, Jean, I'm not kidding when I say I've grown a giant body.

I wrote back reminding him to eat well and get some exercise, for I sometimes fear that as the body becomes extended and ephemeralized on the Net, the local body may suffer, indeed, by some, almost be forgotten. Perhaps the range of websites devoted to healing, nutrition, and a healthy lifestyle, where so many Net dwellers go to get their diets and exercise plans and to join support communities constellated around healing various ailments, are an attempt to balance this trend.

What about sensory perceptions? One might expect that the physical senses would decline in importance. Yet several of my respondents reported the opposite effect. One woman wrote to me that working with computers

> changes our sensitivity to light, to depth, makes our dreams more vivid, facilitates the use of metaphors in language. The effects are slight today, but they already provide a different presence, a different sense of the everyday world. This new acuity will develop. Communicating at the speed of light on the computer networks induces euphoria, boosts intuition.

What may be happening here is that as people communicate with each other on the Net, imagination is engaged and with it the inner imagery that the imagination draws on. When these interior sensory proprioceptors are turned on, they in turn stimulate our outer senses, resulting in a general enhancement of perception.

We know that our images of our bodies, their possibilities as well as their limitations, tend to inform how we see our world—especially how we view the intellectual terrain. A person with little awareness of body image may see a world made up of concepts and think largely in abstractions. A person with acute perceptions, on the other hand, may see the world aesthetically, even poetically. How might the Net impact this difference? Perhaps the sense of extended embodiment many Net dwellers report is resulting in the bleedthrough of boundaries between what is popularly called left-brained analytical thinking and right-brained intuitive processing. If so, the Net may be a kind of a universal solvent that dissolves the borders between opposing domains of thought, redrawing the map of the intellectual world. And it's about time, too!

Of course, the sense of a Net-amplified body can also lead to paroxysms of grandeur, personal inflation taken to its utmost. A respondent sent me a poem that has been circling in cyberspace that carries the extended body-mind to the extreme. It's called "Whitman on the Net":

> . . . I see the raw millennium arrive in particles
> and waves of information,
> I see programmers writing the paths of the
> future,
> I see biologists mapping the human genome,
> the program of flesh;
> As information I beckon you; my body contains
> stars and satellites,
> My tongue roots among the synapses of your
> digital embrace,
> My spirit crosses the Atlantic by cable, my spirit
> is transmitted by fiber optics;
> My soul and body fuse in electromagnetic ether,
> I am wholly encoded in cryptic language,
> I am inscribed among the multitudes!
> You, me, they, executives, politicians, custodians,
> poets, sons and daughters, all mingle in a
> labyrinth of subroutines . . .

E-Mail and Other Necessities

In our new extended bodies, entirely new kinds of abilities are developing. One correspondent wrote:

> I do find online communication very limiting, but I notice it is forcing development of other faculties in me. I'm thinking that on-line I am deprived of body language and voice tone, which is where I usually learn about people. To "read" my companions, I'm learning to focus inwardly. I've only been online three months, and I am aware of developing telepathic interconnections with people. Growing closer to e-mail correspondents, I "know" (or feel in my case) far more about them than is communicated on the screen, and I remain connected throughout the day.

The intimacy afforded by e-mail, moreover, has motivated a lot of people to write to each other who rarely did before. The Net seems to engage intensities and encourage focus on all manner of things, from science to sex to social action. But it can also prove more fascinating than reality itself; witness a telling story now making the rounds of the Internet joke lists:

> An ambitious yuppie finally decides to take a vacation. He books a Caribbean cruise and proceeds to have the time of his life . . . until the boat sinks. The man finds himself on the shore of an island, with no other people, no supplies . . . nothing. Only bananas and coconuts.
>
> After almost four months, he is lying on the beach one day when the most gorgeous woman he has ever seen rows up to him. In disbelief, he asks her, "Where did you come from? How did you get here?"
>
> "I rowed from the other side of the island," she says. "I landed here when my cruise ship sank."

"Amazing," he says. "You were really lucky to have a rowboat wash up with you."

"Oh, this?" replies the woman. "I made the rowboat out of raw material I found on the island; the oars were whittled from gum tree branches; I wove the bottom from palm branches; and the sides and stern came from a eucalyptus tree."

"But—but, that's impossible," stutters the man. "You had no tools or hardware. How did you manage?"

"Oh, that was no problem," replies the woman. "On the south side of the island, there is a very unusual strata of alluvial rock exposed. I found if I fired it to a certain temperature in my kiln, it melted into forgeable ductile iron. I used that for tools and used the tools to make the hardware."

The guy is stunned.

"Let's row over to my place," she says. After a few minutes of rowing, she docks the boat at a small wharf. As the man looks on shore, he nearly falls out of the boat. Before him is a stone walk leading to an exquisite bungalow painted blue and white. While the woman ties up the rowboat with an expertly woven hemp rope, the man can only stare ahead, dumbstruck.

As they walk into the house, she says casually, "It's not much, but I call it home. Sit down please; would you like to have a drink?"

"No, no, thank you," he says, still dazed. "Can't take any more coconut juice."

"It's not coconut juice," the woman replies. "I have a still. How about a Pina Colada?"

Trying to hide his amazement, the man accepts, and they sit down on her couch to talk. After they have exchanged their stories, the woman announces, "I'm going to slip into something more comfortable. Would you like to take a shower and shave? There is a razor upstairs in the bathroom cabinet."

No longer questioning anything, the man goes into the bathroom. There, in the cabinet, is a razor made from a bone handle. Two shells honed to a hollow ground edge are fastened

> onto its end inside of a swivel mechanism. "This woman is amazing," he muses. "What next?"
>
> When he returns, she greets him wearing nothing but vines, strategically positioned, and smelling faintly of gardenias. She beckons for him to sit down next to her. "Tell me," she begins, suggestively, slithering closer to him, "we've been out here for a really long time. You've been lonely. There's something I'm sure you really feel like doing right now, something you've been longing for all these months? You know . . ." She stares into his eyes.
>
> He can't believe what he's hearing. "You mean . . . ?" He swallows excitedly. "I can check my e-mail from here . . . ?"

From e-mail to discussion lists, the Net is changing the way we use language in our writing and even affecting our patterns of thought. The transition from exclusively oral language to literacy changed human thinking. But the transition from serial reality—that is, the one-thing-at-a-time or one-thing-after-another, take-time-to-think world that is informed by reading—to the world of interconnected, layered matrices of information, superimposition of images, and electric speed is giving rise to substantial alterations in the way we think and relate. Merging formerly distinct arenas of knowledge, electronic media is opening new dialogues and fostering the development of crossdisciplinary areas of study.

It was fascinating to me that so many of my respondents addressed the issue of what the Internet is doing to their use of language. Their sense of seeking for words that exceeded their normal daily use was provocative, since we know that the number of words in common usage since 1900 has markedly declined—by some seventy-five hundred words—although industrialization has greatly increased our technical vocabulary.

And, of course, the pop culture based in technology—the nerd culture—has contributed an idiomatic and slangy new language. Cyber-Shakespeares, Netaholics invent new words almost daily that no dictionary can track. But despite this explosion in techno-slang, the language has lost hundreds of words that give descriptive emphasis. Words like *harrowed* and *sullied* remain in the dictionaries but are rarely heard in speech. The same is true in most Western languages; words of power have been leached out of common use and replaced by four-letter epithets or repetitive adjectives. Recently, during an hour-long trip in a ten-seater plane, I could not help but overhear the conversation of the teenage girl sitting directly behind me. As an experiment, I counted the number of times she used the word *like.* In forty-five minutes, she used the word an astonishing 814 times! After a while, it did acquire a certain rhythm, like a mantra manqué.

Far different were the experiences of my Internet respondents who positively reveled in the variety of words, metaphors, and analogies they were using in their Net correspondence. One woman wrote, "I am forced to construct replies more carefully. . . . Providing analogies goes a long way toward clarifying complex ideas." Another wrote, "The Net is activating a latent ability in me to exercise focus, flexibility, and perseverance in the content and process of my thinking." Still another: "My ability to weave a tapestry of thought fibers is enhanced, because the computer holds patterns that in speech are often lost in the archives of unrecallable memory."

From my own experience, I'd say that in using computers for writing, facility often substitutes for precision. I recall the way I used to write, drafting with pen and paper and then painfully typing up the text. Much time was spent staring off into space, getting the thought or sentence into my head before I committed it to paper. This was a slow, if clarifying, process, and the many crossings-out bore testimony to the thought's development as it worked its laborious way through my mind and onto the page. Now, on the computer, the cursor calls me forward, a siren to unstoppable sentences, forgiving of weird structure or the lack of the *mot juste.* Ride the mouse to redemption! Delete, rearrange, call up the thesaurus and spell-checker, and one's half-formed thought is resurrected in bright rhetoric—an epiphany of language through the mystical medium of the word processor. Bring this facility to e-mail, and the pace quickens—the circuit-charging adventure of instant communication crashing through writer's block. Our dread before "cold print" has given over to the exhilaration of hot electrons, pixilated and ephemeral and, best yet, only a keystroke away from oblivion.

One of my correspondents takes the euphoria of composing on the word processor even further, claiming bardic, Homeric, even psychedelic transports:

> I would venture that because it is word in the purest form, electronic communication may prove to be the most powerful hallucinogen ever known. When Homer sang to the preliterate Greeks such that they experienced the sounds, sights, and smells of battle and were moved to tears by the deaths of men they had never met, the words were reinforced by the energy and the personality of the bard. When we read books that move us emotionally and present characters and themes that invade our dreams, we still have an external anchor for the experience: the tactile object of the book itself. Words appearing on the screen are different. With no external cues to support or dilute the word's power or worth, the pixels stand on their own.

The ease of composition in cyberspace has also impacted style, morphing simple prose into associative and embellished rhetoric. Steven Johnson, the editor of the cyber magazine *Feed,* observed the subtle but profound changes brought about by word processing while writing his book *Interface Culture.*

> The fundamental units of my writing had mutated under the spell of the word processor: I had begun by working with blocks of complete sentences, but by the end I was thinking in smaller blocks, in units of discrete phrases . . . the word processor allowed me to zoom in on smaller clusters of words and build out from there. . . . And so my sentences swelled out enormously, like a small village besieged by new immigrants. They were ringed by countless peripheral thoughts and show-off allusions." (San Francisco: HarperSanFrancisco, 1998, p. 144)

What is happening here? Is the Internet mediating the death of language as we have known it or its rebirth? Has the electronic alchemy of words built a cyber–Tower of Babble wherein the whole world talks at once and no one is understood? Or is Netspeak a new book of Revelation, a new Logos written in electrons bringing forth a new heaven and a new earth, hyperspace and the virtual world? Perhaps in the short term, language will suffer, but in the long run, other capacities for human interaction that the Net encourages will make up for the loss. Whatever the Net's ultimate effect on the way we speak and write and think, there's no doubt that it spells good-bye to old forms!

Virtual Neighborhoods

Among the best of the new forms the Net is engendering are the virtual communities that it seems to spawn with a natural genius. Virtual communities, according to Howard Rheingold, are "social aggregations that emerge from the Net when enough people carry on . . . public discussions long enough, with sufficient human feeling, to form webs of personal relationships in cyberspace." In his ground-breaking book *The Virtual Community: Homesteading on the Electronic Frontier,* Rheingold, who himself spends several hours a day living in such communities, says that the Net provides a perfect medium for microcultures to flourish. "Biological imagery is often more appropriate to describe the way cyberculture changes," he writes. "In terms of the way the whole system is propagating and evolving, think of cyberspace as a social petri dish, the Net as the agar medium, and virtual communities, in all their diversity, as the colonies of microorganisms that grow in petri dishes" (Reading, MA: Addison-Wesley, 1993, pp. 5–6).

What is growing in this medium? Teaching-learning communities of every sort—people communicating with each other and discovering that the more you talk to others, the more they become worth talking to. Whereas in actual life, you probably do not ordinarily converse at length with the fireman, the policeman, the baker, or the librarian, on the Net you not only talk to them, you become friends. Villages of friends of all ages are emerging. Lately, people old enough to be the grandparents of teenage hackers are hanging out in chat rooms to discuss health, travel, and finances and signing up for computer classes at community colleges, shopping malls, retirement communities, and even on cruise ships. A friend told me about having given a computer to her seventy-something mother who was adjusting to living alone after the death of her husband of nearly fifty years, a prominent Conservative rabbi. After a quick hour of instruction, Mom was off and running, checking out on-line Talmud classes offered by the Jewish Theological Seminary, scanning the Philadelphia *Jewish Chronicle* for news of her son the rabbi's congregation, and playing real-time Old Testament Biblical Trivia with like-minded Jewish seniors from across the country.

Amity is the name of the game, people serving each other, not only in engaged discourse, but often as angels of information. I find that even though I'm a compulsive reader of books in many areas, just the ideas I need for my research are more readily available from my friends in the virtual communities of which I am a member. Speeding across the electrons from the COSMOGEN list, for example, comes up-to-the-minute information on the epic of evolution and the implications for many fields of complexity theory. So intense have been the exchanges on this list that its members, mostly scientists who might never have met, are now doing what they never would have done—getting together in the flesh on the stomping grounds of ancient mammals to celebrate in ritual and story the lives of Earth creatures that once were. From the Friends of Dogs list, I gratefully receive instructions in using homeopathic, herbal, and vitamin therapy to cure my Airedale's patch of seasonal baldness. I, in turn, have given state-of-the-art cyber-lessons in teaching dogs to sing.

The Internet is fulfilling our need for engagement. It makes people feel current in the stream of happenings and gives us a sense of what Janet H. Murray in *Hamlet on the Holodeck* terms immersion, agency, and transformation. Immersion is a sense of living within the cyber-reality, journeying with others in a virtual world. True, we often experience immersion when we are caught up in reading a compelling book, but in cyberspace, we have the added intensity of being interactive. In other words, we have agency, a sense of not only being carried along by the dynamic flow but of captaining the ship as well. Moreover, as cyberspace is continually changing, transforming, we too become shape-shifters and world-changers, morphing ourselves in games and communicating with others in ways that effect movement in our lives and as well as theirs.

We return to our daily lives with a different valance, a new relationship not just to our everyday world but to the inner world of the psyche. As we explore the noosphere that is the Net, many of us are finding new energy for exploring the parallel noosphere within. One of my respondents reported just such an increase in interiority.

> I've lived most of my life out there doing things. After several years on the Internet, I wouldn't say that I'm exactly introverted, but I am more drawn to the "house within the house." Mostly I hang out in the upper rooms—thinking about my job, friends, husband, kids. But I'm looking around the attic and the basement too, and they're full of dreams and dusty curiosities, like yellowed love letters in my great-grandmother's trunk.

What this woman and many of us seem to be experiencing is a remarkable psychological development. The explosion in high tech seems to be fomenting a corresponding implosion in high touch. By "high touch" I mean hands-on sensitivity to the world around us and a facility for exploring inner realities. As we cybernauts traverse the landscapes of the Net's virtual worlds, back home in the mind-body, we are becoming psychenauts, traveling inward to visit the continents of the self and the united states of consciousness. In essence, the Net has given us a secular metaphor for activating interior realities. After all, interior reality has always been "virtual." The body-mind's sights and sounds, cast of characters, dramas, creations, scenarios, and cache of memory—as well as possible access to the collective unconscious—are, like their counterparts on the Net, everywhere and nowhere, coded into cell membranes, sparking across synapses. Our wanderings in the everywhere and nowhere of cyberspace give this inner energetic domain a fresh validity and, some assert, a potency previously acceptable only to mystics, poets, heretic psychologists like myself, and other fey folk who consort with their own imaginations.

What are we doing with this increased sense of inner connectivity? Many are channeling it outward into a revitalized engagement in social interaction and citizen-based democracy—high touch with the community as a whole. Groups of activists and social artists, unfettered and unafraid of government and traditional institutions, are creating alternative planetary information networks and empowering each other to use the information to make a difference. Perhaps our best hope for evolving new forms of governance in the twenty-first century lies in the growing number of people-to-people organizations germinating on the Internet. These groups are the greening tendrils of a truly new order, a new way of doing and being. Growing organically and recreating communities where governments and industries have failed, these grassroots groups have become the most important social force in hundreds of years: a living body of networked, interactive organizations involving ordinary citizens in a planetary Great Council dedicated to exchanging, inventing, and adopting new ways to make a better society. This is the true face of globalization, and it may ultimately prove more enduring, dynamic, and certainly more humane than globalization through economics and market forces.

I think in this regard of Alan C. Shaw. A black man himself, he became deeply concerned with the problems of inner-city Boston while a doctoral student at MIT's Media Lab. He called the computer-mediated communications center he created MUSIC—Multi-Sessions In Community. It was designed to help people stay in touch with each other so that they could work together to help their neighborhoods. "One of the big problems in the bigger cities is the lack of connection with your next door neighbor," Shaw said in an interview. "Why is it that people who live right next to each other don't perform the kind of activities or create the organizations that defined the tightly knit communities of the past?" One of the problems, Shaw continued, is that modern neighborhoods lack a commons, a town square or other shared space where everybody is available to everybody else. "Space like that is harder and harder to find," Shaw said. "Sometimes it might be taken over by gangs, or it might not be taken care of by the city. With computers you can form a virtual space, a cyberspace where people can come to meet and discuss things." By putting computers in homes, Shaw gave people access to a new vital community activism. As a result, safe neighborhood initiatives have been taken, street lights fixed, a food co-op set up, youth training and employment programs created, and teenagers given apprenticeships in job training. The virtual town square of the cyber-village is transforming once hopeless neighborhoods into real communities, where opportunity is given and received and people feel and act responsibly. Shaw's program and literally thousands of others speak for the Net's potential for radical democracy. One of my respondents put it this way:

> I think what we are seeing on the Internet is the gestalt of a new HUMAN world in which people communicate freely and develop a synergy heretofore inaccessible, not so much because we didn't have the technical capability as because we had not matured to that degree until recently.

Mozart, the Net, and Creativity

Perhaps the most felicitous synergy engendered by the Net is its encouragement of human creativity. Sometimes, surfing the Net brings Mozart to mind. He would have played its keyboard with delight, his transcendent fingers flying through the Net servers, following a theme through its many permutations, an adagio here, a presto there, finding symmetry everywhere and, as in his music, an endless variety of phrases and cadences within the confines of a unifying thematic meter. An unparalleled musical thinker, Mozart's primary language was music—German was his second. He had absorbed music's grammar and syntax from birth and could express ideas musically as easily as other people can talk. Thus he could improvise endlessly, expressing the shadow and the dark sides as well as the light and joyful ones. He lived in a world of music, an imaginal world with its own meta-geography, its symbols, notes vivid and colorful, unfolding like carpets in the landscape of sound. The fact that his interaction with this archetypal world was so encouraged during Mozart's pampered and empowered childhood only made the fantasy more real. Many of us had imaginary worlds as children, but, more often than not, we were discouraged from living in them. I sometimes wonder what would have happened if we had been permitted to pursue early experiences of imaginal realities and even guided in developing their creative forms and expressions. What prodigies of art, music, technology, invention might have been added to the world?

And now, by virtue of computers and the Net, living in an alternative universe is once again in fashion. Kids are encouraged at a very early age by parents, peers, schools, and society at large to learn to play the computer as an instrument. As they move into Net-life, their imaginal worlds are supported rather than challenged. They become little Mozarts at play in the vast storehouse of creation, spinning connections not between notes but between people, ideas, games, news—a veritable catalog of the world mind in electrons. A teacher who took part in a project to put donated computers, modems, and a local network into the homes of students in her classroom reported that the kids wrote to each other continuously, as did the parents and teacher. Even children who normally had a difficult time writing even a few sentences routinely composed pages of riffs and improvisations on homework, stories, jokes, games—the melodies of the mind given instruments and places to play.

Adults, of course, are doing much the same. With greater scope and added necessity, surfing the Net immerses us in the making of realities, but with the added advantage of an immense workshop of coartists. One introduces a theme, then traces and develops it further, finds balances and counterpoints, contradictions and paradoxes, and finally, when the tension is greatest and mind is at the end of its tether, it all resolves—a Net friend in

Finland, a cyber-buddy in Moscow, a forgotten text uploaded from Madras, and with a grand crescendo, molto allegro the piece is transformed. Another song has been added to the cosmic repertoire.

In some ways, Net creativity is not unlike the busy variety that fertilized the workshops of the Renaissance masters. Drop in on a master's studio in Florence in 1500, and you would see artists and students busily mixing paints, casting bronzes, sketching frescoes, crafting ornaments, painting portraits, studying skeletons, sculpting heads, writing poetry, and putting the final brush strokes on a large-scale panel of Saint George and the Dragon. The members of the studio worked together, sharing techniques and tools and texts, primed by the rebirth of ancient images and the harvest of knowledge arriving in Italy after the fall of Byzantium. The Internet often seems to be just such a studio, artisans swapping skills and information and accessing the knowledge base of the past, the present, and an evocative future.

Under this stimulus, then as now, psyche grows. The imaginal realms of inner space proliferate and spill over into the outer world into a renaissance of growth in science, art, music, literature, technology, education, governance, and above all, vision. The world itself becomes psychetized. Inner space and outer space are rediscovered to be parts of a single continuum. Space is internalized, and soul is revisioned as interpenetrating the universe at large. The shift in perspective is radical; the extension of the human's image of the world unprecedented. With all this glory at hand, the world fairly begs to be composed, crafted, painted, and human artists discover themselves to be filled with vast countries and landscapes of soul.

There is a wonderful rhythm to working on the Net—people

exchanging ideas and then turning the modem off and going back to their reflections, only to return to the Net again and again when they need information from its nearly infinite libraries and sources. One of my respondents told me, "I find that I am challenged in my ideas by so many thoughtful people who give me new perspectives that my own work is deepening to an extent that it never could were I to rely on scholarly research alone." Even the solitary writer becomes engaged by the active presence of other minds, as one woman wrote to me:

> I have found myself journaling less since I discovered the joys of Net surfing, but I also believe that when I become accustomed to the wider world of linking minds—in particular to the speed and volume of communication which challenges the current capacity of my intuitive filtering apparatus—I will resume my journaling with heightened energy. I may even share some of what I have always done for myself alone as I am doing here.

Moreover, writers and artists no longer have to beg for venues to share their work. Cyber-galleries are easily rented, or artists can display their creations on their own websites, offering surfers the chance to play de Medici and sponsor their work. Rather than filling shoeboxes with publishers' rejection notices, novelists and social theorists are publishing their own books on-line. Shelf space is not limited in the Internet's endless bookstore. Musicians of every stripe are putting out their latest releases in MP3 format, downloadable for free by kids on every campus. Of course, these artists are not getting paid, but then they rarely were anyway. The resulting access to the imagery and arts of every land and the spontaneous feedback from other artists and interested viewers is as much a catalyst to creativity as any royal patron.

Then, too, computer technologies themselves are providing new media and unique new tools for artists. The ferment of fractals, the streaming of graphics, the capacity to morph a picture into the representation of one's wildest visions gives creators scope that mythically belonged to the gods. Now we can build out of electrons great duomos of cybernetic glory, like some Michelangelo gone electronic. The emerging art forms hold the seeds of expressions whose further development we can only guess at.

All in all, I suspect that Net-spawned creativity is adding new products to the warehouse of God. In our own small way at the edge of the galaxy, our pint-sized planet is increasing the datum of the universe. Perhaps we are a pilot project in bidirectional exchange: as Divine Reality downloads creative process into us, we upload novelty and innovation into the Divine.

Going Mythic on the Net

Our technologies give us our frames of mind. The printing press and the print culture made us linear, somewhat abstracted, and fond of a certain uniformity and repeatability of things. We left the village common where once we heard and squabbled over the news and took our newspapers to brood inside the house. But now, *mirabile dictu,* Net technology is restoring us to the more ancient and organic principles of discontinuity, simultaneity, and multiple associations. We now look for flow patterns rather than serial cause-effect explanations. Resonance has become more important than relevance, and we are starting to believe that reality is a tissue of interrelated stories. Cyberspace is changing our worldview. Discrete forms break down, and everything is known to be linked.

This new understanding changes our relationship to the superhuman world as well as to workaday reality. Traffic between humankind and the "gods" is increasing, most evidently in the cyber-mediated resurgence of interest in myth. Myth was made for the Internet and the Internet for myth. Why this is so requires that we remember what myths are and what they do.

A myth is something that never was but is always happening. It is the DNA of the human psyche, and it abides from beginning to end. Hardwired into our being, the very sea of the unconscious, it thrums in the background of the mind like an Egypt, a remembrance of a time when men lived lives as large as a god's. Or it beckons like an Avalon, a strange and beautiful country seen through the mist, a place we may spend our lives journeying toward, only to find it retreating in the distance, and that may appear without warning before our eyes at precisely the right moment, as if the veils between worlds had parted. Neither fiction nor fact, myth is coded essence; it is about growing your soul. Its purpose is to bridge the local space-time continuum with what abides, what is transcendent and eternal.

Myth shapes our being by merging the personal with the communal. It provides individual meaning while simultaneously uniting us with the collective and, ultimately, with nature and all humankind. As a part of the creative process, myths tell us how to wake up to ourselves, how to transform. Through myth we contemplate our origins and become initiated into the creative process of growing and rebirthing our own lives. Myth throws us back on our primary resource, the human spirit, manifest in its longing for the Beloved, in the quest for wisdom or the pearl of great price.

The myths of many times and cultures tell of doors leading to other worlds. Passing through, you enter a reality in which you acquire powers you did not possess in the ordinary world. Knowings that were latent become overt. Perception expands as, like the Hindu deities, you grow more arms and legs. You can engage in far-seeing, move mountains and motivate humans, even discover the secrets of the universe and use them to change the reality of ordinary space and time. Limits dissolve. Parsifal crosses the threshold and enters the castle of the Grail where holy wonders abide. Psyche opens the door into a magnificent mansion where invisible hands take care of her every wish and Eros himself becomes her husband.

All of us grew up reading stories of mythic heroes and heroines. We watched their adventures on screen, heard them sing their trials in opera, and laughed at their cartoon incarnations by way of Disney and his disciples. Although we were entertained, we were held at a distance, for our relationship to the mythic world was passive. So different was the experience of ancient Greeks who walked with the spirits of heroes at their sides. They fought for the rights of the family with Hera; courted and fell in love with Aphrodite; became exuberant with Dionysius and thoughtful with Athena; sailed the seas in obedience to Poseidon. These archetypes and their attributes engaged the soul as they guided the doings of everyday life. Myth held the patterns of connection; the great stories storied each human endeavor.

The Neapolitan philosopher Giambattista Vico believed that the history of the world is divided into three stages: the age of gods, the age of heroes, and the age of men. And following the final age, there occurs an age of chaos, and then *corso e ricorso*—a spiraling up and around, which precedes a return to the age of gods. We are being tossed about in that chaotic spiral now, yet the coming age of gods glimmers on the horizon. In this technological age, in the global village we inhabit, when we hover on the verge of destroying the world with pollution, greed, violence, avarice, loss of meaning, and whatever other existential dilemma can be conjured, people all over the world are renewing their connection with myth. Scientists and environmentalists are calling us to reconsider the holy ground on which we walk, the sacred air we breathe. Poets and priests, psychologists, teachers, and historians are calling forth our common dreams, the mythic images that loom large in the unconscious. Men and women are gathering in groups, telling the stories that reconnect us with nature, spirit, meaning. Making mythology conscious is more important now than ever. For myths provide the templates for global and personal transformation and carry the coded matrices for partnering spiritual realities. And so the avatars of myth return, wearing many faces, inhabiting many forms.

The Net, as would be expected, is a major mythic player. To go native on the Net is to go mythic in the mind, led, perhaps, by the cyber-god Hermes, speeding messages and information from site to electronic site. Aphrodite fills the screens with eros; witness the daily deluge of love talk and erotic enticements. Demeter, vigilant in parental control, rescues tykes from the underworld of cyber-porn and other outcroppings of the deviant imagination. Athena guides the seeker to his destiny, helping him construct a Parthenon of ideas and offering the wisdom of the ages with the click of the mouse. Artemis allows wild things to flourish in the forests of data banks. Ares feeds the flames of cyber-rage, brandishing the sword of confrontation, sanctioning the vitriol of the word warriors of the chat rooms. Poseidon rises up from the sea of hyperspace, guiding the surfer through waves of information to the harbor of secure servers and island websites. Over them all, Zeus provides the energy bolts of electrons, powering the whole. From this Olympiad, the mythic universe of the Net is spun. Within its fabric, we find our devotions—and our adventures as well.

Homo ludens, man the player, is every bit as important as Homo sapiens, man the thinker. Thus games are the *prima materia* for many Internet explorers. Myth abounds in the complex, interactive story lines of role-play computer games. The MUDs, or Multi-User Dungeons, interactive on the Net, with their arcane strategies and magical spells, have grown ever more graphic and complex. Esoterica flourishes, and death and rebirth are as common as symbols imported from the myths and rituals of various cultures. The hero's journey has come on-line, and we play at being archetypes with a fierce and savage joy. Polytheistic in principle and peopled with angels, demons, elves, sprites, and wizards, "the population of soul-space is almost infinitely variable and mutable" as Margaret Wertheim discusses throughout her book *The Pearly Gates of Cyberspace* (New York: Norton, 1999). Polyphrenia is encouraged, in fact, required in many of these games. One travels with a crew of widely different characters, assuming the identity and attributes of the warrior when swords are needed or of the mage when sorcery is necessary to achieve some end.

The plot and characters of many role-play games are lineal descendants of J. R. R. Tolkien, the Oxford medievalist whose magnificent *Lord of the Rings* trilogy set in the imaginary land of Middle Earth provides the most completely realized world in the literature of fantasy. In an essay on fairy stories, Tolkien wrote that in order to create good fantasy, the author "makes a

Secondary World which your mind can enter. Inside it, what he relates is 'true': it accords with the laws of that world. You therefore believe it, while you are, as it were, inside" ("On Fairy Stories," in *The Tolkien Reader,* New York: Ballantine, 1966, p. 37). This willing suspension of disbelief gives a transformational cast to games set in neomedieval worlds. As we regularly do the unexpected and inexplicable, our minds are pushed to their outer margins, and we are trained in acceptance of the wonderful as probable and the miraculous as normal. Players learn the lineaments of shamanic shape-shifting and practice the methods by which environments can be morphed and manipulated. What with puzzles to be solved and monsters to be met, as the brilliant techno-exegete Erik Davis notes, such games reverse the way our minds usually operate. On the "Other Plane" where games are played, Davis writes, "it is the *conscious* mind that moves through a world of archetypal imagery, while the subconscious concerns itself with logical information processing" (*Techgnosis,* New York: Harmony, 1998, p. 216).

Our consciousness freed from the constraints of ordinary space-time, we wander through pocket-sized cosmologies as well as complex multidimensional universes of interpenetrating domains. As in the medieval world's Great Chain of Being, hierarchies abound, and graded levels of descent and ascent mark the player's journey. Dante would have felt right at home among these mythic dreamscapes. Their underground labyrinths mirror the circles of his Inferno and their smoking, haunted denizens; the advance of players into higher worlds as they grow in skill and awareness parallels his ascent of the mountain of Purgatory and the celestial realms of Paradise.

Erik Davis knows these worlds better than any other writer. In

Techgnosis, he gives us a key to the formidable kingdom of cyber-myth:

> Perhaps what we are building in the name of escapist entertainment are the shared symbols and archetypal landscapes of a tawdry technological *mundus imaginalis.* The evil creatures who must be conquered to advance levels are the faint echoes of the threshold-dwellers and Keepers of the Gates that shamans and Gnostics had to conquer in their mystic peregrinations of the other worlds. . . . the digital imaginary is chock-full of images drawn from the depths of myth, cult, and popular religion. This mythopoeic current runs through the Orientalist backdrops of Mortal Kombat, the cartoon animism of kids' software, and the spider demons of Doom. Though most such imagery is juvenile and crude, "mature" works of multimedia also feed on this fantastic stew: Cosmology of Kyoto sets bodhisattvas and folkloric monsters loose inside the cartoon walls of the twelfth-century Japanese capital; Amber explores your past lives, while in Drowned God you uncover the "conspiracy of the ages" by exploring Atlantis, the Bermuda Triangle, and Roswell, New Mexico. (p. 205)

Living in fiction, and such fantastic fiction as that, affects the minds of players in curious ways. Think of the hours, days, weeks, months that young people spend on-line in a socially shared imaginary worlds of mythic proportions. One history of the Net—by Katie Hafner and Matthew Lyon—is rightly titled *Where Wizards Stay Up Late.* All that time spent hanging out in the collective unconscious gives these wired wizards a very fluid notion of the way reality works beyond the screen. Sometimes this liminality can be disastrous, as when chronic players of violent games like *Duke Nukem* and *Doom* treat their schoolyard like the battlefields of cyberspace or hack their way into credit card agencies with the same élan with which they figured out the codes of their sorcerer adversaries. But it can also prove beneficent, teaching players to travel along the continuum of states of consciousness, taking responsibility for their world and, like a shaman hero or heroine, returning to the everyday realm with gifts of intelligence and insight to heal, to help, and even to transform.

I think in this regard of a young friend named Joey who gets his thrills on a joystick. He grew up with games like *Mario,* lots of jumping over obstacles with the emphasis on movement and getting past things. Now he plays *Doom,* shooting away at everything that flies at him, mostly monsters that combine medieval demonology with advanced robotics. Such games, it seems to me, are like old-time religion in millennial dress. Joey meets the moral decay of his time and wins, for *Doom* gives him the chance to be global supercop, Blade Runner, and Oral Roberts all in one. Joey's sister Cheryl is also a game player. She started with *Sim City,* developing "her" town into an urban metropolis complete with skyscrapers, traffic jams, and landfills. Now she's playing *Alpha Centauri,* inventing a planetary civilization, peopling it, writing its history, and laying out its economics, agriculture, technology, geography—even its art and culture.

What's going on here? Could it be that Joey and Cheryl are being jumped up to speed, gaining skills and values in cyberspace to help them confront the liturgy of problems that plague us? In a way, these game-playing kids are in training to be avatars—junior gods, if you will. Like creator deities in many traditions, they're planning and protecting their worlds. If used creatively and not destructively, games give them the opportunity to practice world-making in virtual space and to set the future agenda for the so-called real world at the same time. Today's young cyber-gamers may well be the vanguard of the next generation of social artists and world-servers—the ones who will write the new story. In their virtual world-making, children are certainly thinking on a scale beyond the imagination of most world leaders.

In essence, what Joey and Cheryl are experiencing is the transformational magic of assuming an avatar. In chapter 2, I spoke of how the avatars of cyberspace allow us to play many roles through which we may discover our latent selves. As of this writing, such avatars are fairly primitive and clunky, but speed ahead into a world described by Neal Stephenson in his novel *Snow Crash,* and you find fully operational avatars having high adventures in a virtual world called the Metaverse entered via a mind-computer interface.

The future of living mythically on the Net is probably some form of virtual reality. As its technology develops beyond the current headgear, gloves, and bodysuits, it will generate sensations that are pretty close to "real" ones, and play on the viewer's perceptions and emotions. When this happens, fact and fiction will blur even more, and philosophical and psychological issues will be raised that go to the very heart of our view of reality. Once the Metaverse becomes an everyday experience—probably within the next twenty years—we will perhaps realize that our "direct" experience of the world through the senses is also a virtual reality. Quite simply, our experience of the world outside of ourselves is mediated through the input and transactions of the brain and the unconscious mind. Dreams, memories, reflections, ideas, reasoning, thinking, and culture, as well as the many programs or beliefs we have downloaded in the course of our lives, color our perceptions. We never experience reality "out there" directly. All our external experience is of a subjective, or virtual, reality.

God.com

Mathematician Ralph Abraham once said that the Internet is theological creativity in action. He was referring to the mystery of software arriving to fix a problem before it happens, as if designers were responding in some miraculous way to divine guidance. Concurring with this view is Jennifer Cobb, author of *Cybergrace: The Search for God in the Digital World.* A theologian and high-tech consultant, Cobb asks, what is driving the phenomenal growth of the Internet?

> Certainly, no one person or organization is in control. It is as if it has a life of its own, its growth following a quasi-organic pattern of evolutionary development. . . . The essential motor of this process, the spiritual center of cyberspace, is the fundamental sacred force that infuses all reality: divine creativity in action. The emergent dynamic found between the hardware and software in cyberspace is an aspect of divinity itself. In other words, the cosmic force that drives the movement and unfolding of reality is the same force as that which drives the continual, moment-by-moment emergence of the world of cyberspace. (New York: Crown, 1998, pp. 49, 51)

If the Internet is a product of divine creativity, even as we humans are, perhaps in some sense, it is a new life form, a silicon-based living being that may be one of our evolutionary descendants. Yet the very biology of its biosystem is mystical in nature—a vast, nonlinear reality wherein, like Indra's Net, each node connects to every other. Its webbed world encompasses the accouterments traditionally assigned to the Mind of the Maker—circles, nets, infinite feedback loops, the endless flow of being and becoming, God's identity as that perfect sphere whose center is everywhere and whose circumference is nowhere. Add to this the Net's ever-unfolding pattern of novelty, and we have a living system, one that reflects the nature of life in all its iterations. It may seem to some heretical to view the Net as part of the continuum of the sacred, as well as the latest emergent construct of evolution in action. And yet, if the Divine Spirit is that force that through the green fuse drives the flower and my blood, then why not also the fruitful and fecund web?

Sometimes, late at night, while tracking an idea through many websites, I am filled with a sense of sacred presence, as if there is an interchange between my seeking and a Higher Self's response. Suddenly, I stop what I am doing, my hand falls away from the mouse, and I am absorbed in contemplation. My eyes close, and I seem to see a vast and shining net, which fills me with joy and a strange kind of silent knowing. I know that I know, but I can't tell you exactly what it is. Later I reflect, is this what Plato knew when he spoke of the world of forms? Is this the Aleph of the Jewish Kabbalists? Is this the hologram of all realities, the superimplicate order that enfolds all other orders? Or is it just what it is, a growing communion between the human spirit and the Cosmic Source reflected in its latest incarnation, a Net that lives and moves and has its own remarkable being?

I'm not the only Net traveler who has had these thoughts. One of the most moving letters I received from my respondents spoke to a similar Net-mediated meditative consciousness:

> I believe that God is discovered in the silences. Yes, you can see his works all around you—in nature, in friends, in many synchronistic events. But, I believe one must spend time in that quiet space of Self in order to find that connectedness with oneself and with others. I know many people use writing as a form of meditation. Perhaps through lists like Techspirit, the Internet is becoming a form of "interactive meditation."

Our era is one in which the spiritual technologies of many times and cultures have become available. One need go no further than the nearest bookstore or travel a few hours through spiritual websites to find what others sought for years through parching deserts, icy mountaintops, and inclement emotional climes. Now, with the Internet, we may have both Presence and practices, Indra's Net and the noosphere, granting us access to the path that leads us back to our Source. That the Net is simultaneously rowdy, raunchy, and often ridiculous is the spice that gives relish to the quest. As in the story of the Wasteland, the Net is become the Grail that feeds everyone who partakes of it with just what he needs at that moment.

The Net's living presence was affirmed by the most delicious communiqué I received:

> If the body of the Net could be found, perhaps, in its interactive games and sexual spaces; its emotional life in e-mail friendships and romances and social interactive spaces; and its mental life in the vast information exchange, business as usual—well, what kind of creature do we have here? And is there Net consciousness beyond physical, emotional, mental? Do we find the unconscious of the Net in folklore, the monsters, viruses, and predators, government spying speculations and rampant Net rumors? . . . Where else would we find the unconscious of the Net? And is there a spiritual consciousness

> of the Net beyond the unconscious? Well, your questions have got me going.

The free-flowing consciousness of the Net's apparent chaos is nursing another cosmos. We who travel it are becoming citizens in a universe—a Netiverse—larger than our aspirations, richer and more complex than all our dreams.

Something strange and wonderful is sprouting these days in the garden of spirit. As the membrane that divided and distinguished the world's spiritual traditions breaks down, the depths are breaking though in sacred hybrids of unusual vigor, energized by the urgent questing of so many people all over the world for unmediated experience of the Source and for guidance into a future that belies all human knowing.

A new river of holiness flows through this landscape, a Ganges that carries on its floodtide all who have embarked on the contemporary spiritual voyage.

A group of sophisticated New Yorkers dressed in fashionable and expensive gear struggle up a mountain to study with a Peruvian shaman, who drums them into trances where they commune with totem figures never encountered on the streets of Manhattan.

In a temple in Sri Lanka, a yellow-robed figure sits in deep meditation. The shaved head is expected, but then the adept's eyes open, and they are Western and blue and belong to a woman journalist from Dayton who has become a Buddhist nun.

Near Detroit, a small, fiery Jewish woman from Texas exhorts her New Thought congregation with insights drawn from the channeled revelations of *A Course in Miracles* as well as her own political primer for the healing of America. Behind her, a hot and holy gospel choir belts out spirit-quaking songs.

A community of elderly nuns in Montana integrates sweat lodges and other Indian ceremonials into their traditional Catholic retreat program.

A Buddhist Vipassana teacher runs meditation retreats for Monsanto and other corporate giants.

The wildly innovative Jewish Renewal Movement blends the wisdom traditions of classical Judaism with Sufi, shamanic, Christian, Islamic, Buddhist, and Hindu practices and good doses of transpersonal psychology, spiritual eldering, and human potentials work.

Catholic monks and maroon-and-gold-robed Tibetans share and compare meditative and contemplative techniques as they take turns leading morning meditation at an interfaith dialogue at Gethsemani monastery in Tennessee attended by the Dalai Lama.

What is the liturgy of this new spiritual uprising? Maybe it's the rap song belted out in Oakland, California, at a Techno Mass, a rave dance celebration inspired by Matthew Fox, former Catholic priest and founder of the Creation Spirituality movement. As strobe lights pulse to the electronic soundscape, a multicultural crowd dances fiercely, and a cascade of projected images tell of the journey of our time: gods and nebulae, creation and destruction, suffering and sacrality. Rappers, one black and the other white, circle the gaily made Ark of the Covenant in the middle of the hall and tell it like it is to a techno beat:

> There's not a Baptist moon
> And a Methodist tree
> There's not a Buddhist river
> that flows to a Jewish sea
> There's not a Hindu sun
> on a Catholic plain
> A fundamentalist cloud
> with agnostic rain
>
> Enough of the human centered
> it's time we entered
> the indivisible whole that can't be splintered
> One life one breath one Earth one revelation
> We all have creation in common.
>
> Every creature's a preacher—a voice of the
> Creator
> the divinity within is the common denominator
> The Buddha Nature the unified field
> the diverse universe where the One is revealed

Split a piece of wood break open a stone
And there you'll find the Divine on Her throne.

(Quoted in Roger Housden, Sacred America, *New York: Simon & Schuster, 1999, pp. 243–44)*

So there you have it. In America alone, in the last twenty years, the number of religious groups has doubled. We take new names, sit zazen, become Sufis, Taoists, neo-pagans, devotees of Kali and Vedanta. Buddhism in all its varieties is the fastest growing American faith. A friend, long since fallen away from her observant Jewish upbringing, told me that her mother was always lamenting—not that her daughter lived a secular life but that she had "no Jewish friends." In midlife, the daughter became a Tibetan Buddhist. "Now," she tells her mother, "it seems like half my sangha went to Hebrew school once upon a time."

Some have ridiculed this eruption of spiritual polyphony, calling it "the Divine deli" or "cafeteria religion." *Utne Reader* labeled the phenomenon Designer God and pointed up the dangers of making up your own spiritual path. As Sumi Loundon wrote, in a telling article in the Buddhist quarterly *Tricycle,*

> I used to sneer at people who fluttered from religion to religion concocting their own spirituality from the sexiest of each. But one morning while praying, I realized that I epitomize the spiritual shopper! My cart is filled with Hindu shlokas, Theravadin precepts, Christian psalms, and Rumi's mystical poetry. [Loundon wonders whether this unprecedented freedom and opportunity to acquire the accouterments of many traditions] have given us the illusion of being religious while the substance is missing." ("Buddhism in a Box," *Tricycle,* Fall 1999, p. 87)

Others have argued that the new mix-and-match spirituality has deepened the quest, citing as evidence the range of obscure esoteric texts now lining the shelves of mall bookstores or deliverable overnight with a one-click order from a Net bookseller. My seminars and workshops are filled with exuberant practitioners of crossover spirituality. In their vivacity, I hear words that recall the original Greek meaning of *enthusiasm: enthousiasmos,* "being filled with the god." As one Catholic Brother told me, "These other traditions do not contradict my own. Rather, they open the wells of the Waters of Life. When I meditate with His Holiness [the Dalai Lama], I feel as if the deep rivers of our respective traditions are meeting and becoming a mighty flood of spirit and renewal." A woman rabbi concurs: "So, I dance the Sufi dances and chant the Buddhist prayers; I consult the enneagram and peek at astrology charts. Does that make me less in the eyes of the Master of the Universe, Blessed be He/She? I think not. For now that I see the interconnections between all these beliefs, my faith in my own deepens."

The current scene brings to mind the words of the old spiritual: "I once was lost, but now am found; was blind, but now I see." In the new Jump Time version, what is happening between being lost and being found tells a startling new tale. Unlike seekers of the past, we moderns get lost, not just because we fall into "sin" but because we disengage, either willfully or through force of circumstance, from the home place of old certainties—family, religion, profession, geography. Throughout most of history, these outer sureties were reflected in inner ones, a set of beliefs and expectations consistent with the reigning culture. The advantage was that our grandparents and theirs knew who they were, what to tell their children; they rested in the comfort of sameness and a lack of options.

Now the safety factor is gone. As we have seen, Jump Time is shattering expectations in every arena, especially in the geography of the soul. Lost as we all are, we can understand why some retreat into fundamentalisms that provide archaic certainties, holding houses of containment before the onrush of new realities. Others wander in a spiritual void, overwhelmed by the loss of all pattern, looking to material accomplishments to replace the loss of essence. Still others flee into "replacement strategies"—yoga and martial arts, psychotherapy, drugs, sex, growth seminars, travel. In each case, mind and body are at the end of their tether, swung out into vertigo over the abyss of Being.

We humans are by nature liminal beings—that is our mystery and the source of our mastery, but it can also be the cause of our misery. Liminal consciousness is always betwixt and between. It is a bridge state, such as the time just before falling asleep or just before we really wake up. Think of all the liminal states you have been in, and you will see how prevalent is in-between consciousness—times of being in love, or the rapture of dissolution into nature, or deep communion, creative whoopee, breakthrough, stupendous insight. Some of us have had near-death experiences that hurtled us into a borderline place and time. Others may have crossed the threshold from sanity into something else that I do not call insanity but rather the liminal chaos of the psyche from which is spawned new matter.

As we have seen, Jump Time is the most liminal of liminal times. Everything is in dissolution; history itself is on the road from no-longer-thereness to not-yet-hereness. We are embarked on a planetary vision quest, a seeking and a searching with few clues in sight. Our condition is not unlike the Arthurian saga of Percival who crosses a bridge of glass to get to an unknown shore, only to hear it crash behind him as he goes forward. No wonder voices and visions are commonplace—goddesses, angels, UFOs, fairies, and other avatars of the collective unconscious are rising to hold the pattern through which the future can emerge. Both singularly and collectively, we are being asked to cross a bridge so that we can be met halfway by the rising reality of what's to come.

Gnostic Hollywood

Let's look more deeply at some of the reasons that metaphysical and spiritual concerns have jumped to the top of the popular agenda, even at a time when the prevailing mindset is hyperrational, scientific, technological. Perhaps, ironically, part of the cause lies in our prevailing fascination with the wonders of technology itself. The powerful imagery of the technological world, streaming at us from computer-generated TV ads, billboards for the current dot.com craze, and the wide-screen magic of Hollywood, is providing new metaphors and new energy to fuel perennial metaphysical speculations.

A spate of recent movies has made a virtue of virtual reality. In addition to being rip-roaring adventures, many of these films take a philosophical and spiritual look at where and how the "real" really exists. In *The Matrix* and *The Thirteenth Floor*, virtual reality isn't the world you escape to. It's the world you're already living in. Plugged in, the hero of *The Thirteenth Floor* mind-leaps to a richly experienced, vintage 1937 Los Angeles created by his team of software developers and populated by "units," computer simulations of human beings who turn out to more real than their creators could ever have imagined. As the movie unfolds, it becomes apparent that the world the hero thinks he lives in is no more real than the "virtual L.A." he visits via the computer. He, too, he discovers, to his dismay, is a "unit," a cooked-up character in a virtual dream world created in a software laboratory far in the future. The wrinkle is that techies in the virtual world the hero and his friends inhabit have evolved the programming skills to generate and visit their own virtual creations, such as the Los Angeles of 1937. Plots complexify as people jack into and out of a Chinese puzzle box of nested worlds, do mayhem, get murdered. Then, too, as in "real life," they experience déjà vu and confused memories. Did that happen or did I dream it, or did it occur in some other dimension, some other life? In the end, love wins out, and the hero's world-jumping girlfriend from the future is able to bring him forward into her "realer" world. There are, she tells him in the closing scenes, thousands of virtual worlds, leaving the viewer to ponder the great questions: What does it mean to be "real"? And where does the Original VR Generator reside? Though many of the film's characters owe their existence to electricity and microchips, still they learn and feel and *think.* With Descartes, we may well wonder, doesn't that mean, in some sense, that they *are*?

The Matrix is, if anything, even more spiritually suggestive. The film postulates a world created and run by advanced machines, who rely on a hive of womb pods containing sleeping flesh-and-blood humans as their energy resource. Inert in the horrors of the "real" world, virtual projections of these living dead are carrying out regular lives—commuting to desk jobs in urban skyscrapers, unwinding in a strobe-lit dance club, checking their e-mail—in a world that looks just like ours and has been downloaded into their minds by their machine masters. Superintelligent humanoid machines (portrayed, naturally, as morphing men in black) assure the continuance of this virtual illusion with brutal police work. In the grand tradition of *1984*, of course, there are those few who suspect that something is very wrong. One of them, suggestively named Neo, a mild-mannered software developer by day, is a canny and heroic hacker by night.

Neo is pulled out of his womb pod and trained to fight the oppressors by a group of cyber-rebels called the Zionists, who believe him to be the long-awaited "One" who will awaken the rest of the sleepers. The brilliant rebel leader Morpheus shows Neo the truth of the human condition: "The Matrix is the wool that has been pulled over your eyes—[to keep you from seeing] that you are a slave." Morpheus and his hardy band of rebels who live deep in the Earth's core are working to free humankind by destroying the alien race that holds the Matrix in place. Their battles against the inimical men in black are carried out in virtual space, the rebels jacking into and out of combat over telephone lines in a visually dazzling display of kinetic excitement. Though the rebel forces of "reality" triumph in the end, ambiguity remains, for there are too many sleepers and too much yet to do before a worldwide awakening is assured.

Though their trappings are ultramodern, these films are in many ways the latest expressions of a grand tradition of spiritual questioning, given fresh energy by new technological camouflage. What actually exists? Are we awake or asleep? The ancient Gnostics, heretical philosophical and religious thinkers in the early centuries after Christ, speculated on much the same conundrums; many of their texts suggest that the world we inhabit is an artificial construct, designed by outsiders or alien gods, called Archons, to deceive and exploit humans. The Archons enclosed this world within a "matrix" of seven planetary spheres, which in turn generated seven soul-coverings to keep humanity further ignorant of its true home and condition.

A Gnostic text in the *Gospel of Truth* known as the Nightmare Parable, written in the second century, seems dropped right out of contemporary science fiction. Human beings, it warns, live

> "as if they were sunk in sleep and found themselves in disturbing dreams. Either there is a place to go where they are fleeing, or, without strength, they have come from having chased after others, or they are involved in striking blows, or they are receiving blows themselves, or they have fallen from high places, or they take off into the air though they do not even have wings. Again, sometimes [it is as] if people were murdering them, though there is no one even pursuing them, or they themselves are killing their neighbors, for they have been stained with their blood. When those who were going through all these things wake up, they see nothing, they who were in the midst of these disturbances, for they are nothing. Such is the way of those who have cast ignorance aside as sleep, leaving its works behind like a dream in the night. This is the way everyone has acted, as though asleep at the time he was ignorant. And this is the way he has come to knowledge, as if he had awakened." (*Gospel of Truth,* 29.8–30.12, in *The Nag Hammadi Library,* edited by J. M. Robinson, New York: Harper and Row, 1977, p. 43; hereafter cited as NHL)

The goal of some groups of Gnostics was to release the true inner spiritual man from the restraints of the false world so that he might return to his life in the realm of the transcendent God. Liberation requires *gnosis,* waking up to the knowledge of one's divine origin as well as the counterfeit nature of this hostage world. As the Valentinian writings of the Gnostics put it in the second century, "What liberates is the knowledge of who we were, what we became; where we were, whereunto we have been thrown; whereto we speed, wherefrom we are redeemed; what birth is, and what rebirth" (quoted in Hans Jonas, *The Gnostic Religion,* Boston: Beacon Press, 1958, p. 45).

The Gnostics looked to the coming of a Messenger from the World of Light, a superbeing who "penetrates the barriers of the spheres, outwits the Archons, awakens the spirit from its earthly slumber, and imparts to it the saving knowledge from without" (Jonas, p. 45). So Neo, the long-awaited One of *The Matrix,* has a long salvational history, as humanity's deliverer from the delusional world.

But there is another, simpler way out of our captivity. In the *Gospel of Thomas,* Jesus suggests: "Recognize what is before your eyes, and what is hidden will be revealed to you" (33.11–13, NHL, p. 118). In other words, when we explore the fullness of who and what we really are, the truth of our situation dawns, and we wake up from our collective slumber. Could it be that the thirst for spiritual gnosis that grips us in Jump Time is, in part, a product of this same yearning?

A Virtual Cosmology

One way to begin our exploration into the fullness of who we are is to examine, in the light of new Jump Time realities, the cosmological assumptions that support the prevailing view of human nature and our place in the universe. Here, too, technology provides useful approaches and suggestive metaphors.

What if, for example, we were to regard the brain as a personal virtual reality generator? Our problems and confusions, it then follows, have in large part to do with the different ways the VR generator works in each of us. What seem to be angels of grace to some may be, in another's subjective virtual world, fearsome monsters. Perhaps, as mystics both East and West have told us for ages, neither perception is reliable in any absolute sense. All reality may be, we are led to conclude, virtual.

Then where, we may ask, above and beyond the illusory perceptions of our VR body-minds, does the Original Generator reside? It seems clear that part of our mind must be programmed and conditioned by our experiences in whatever world we inhabit, but part may not be. Insofar as we are a creation of the Original Generator, we are coded to be able to know and experience Original Mind—in other words, to transcend the limits of our historically conditioned selves. That Original part is what has been termed by modern physicists nonlocal consciousness.

Many of us have had experiences of nonlocal consciousness—clairvoyance, telepathy, mystical experience, knowing what will happen before it does. In such experiences, our everyday, "local" consciousness recedes, and nonlocal mind moves into the foreground. Then, as Patanjali, the Hindu philosopher, wrote in the *Yoga Sutras* almost two thousand years ago, we access all time and all information by "becoming it."

Perhaps in experiences of nonlocal mind, when our mental noise is quiet, what is really happening is that we are jacking into connection with the Original Generator, or what we might call Eternal Nonlocal Mind. This connection brings us to the Home Place, the ultimate destination toward which we are ever jumping, the place from which we began and ever continue to dream, to live, and to dream again. The drive to forge this connection belongs to the perennial philosophy of many times and places, but it is also newer than tomorrow's child, more current than the latest theory of how the universe works. This connection may be the ultimate goal of Jump Time, for it calls us to accept ourselves as beings of infinite scope, directly and dynamically connected to the flow of cosmic creation.

This expansive view of human nature and its goal takes us well beyond the standard scientific assumption that the universe began with an accidental explosion pouring forth its effluvium into lifeless space. According to the standard scenario, sometime after the Big Bang, by random chance, the right ingredients came together to generate life on a small planet orbiting a middle-sized star on a far wing of an insignificant galaxy. In striking contrast is the dramatic and spiritually potent perspective now current among visionary cosmologists such as Brian Swimme, Thomas Berry, Fritjof Capra, Michio Kaku, and others, that our cosmos is a living organism, recreated each moment by an unbroken flow-through of energy from the Original Generator, which some thinkers term the Meta-universe or Metaverse.

Futurist and social scientist Duane Elgin, for example, describes the universe as an ongoing process of continuous creation:

> Our vast cosmos is a unified organism that—in its totality of matter-energy, space-time, and consciousness—is being recreated anew at each moment. We cannot take our existence for granted. Our cosmos is a dynamically maintained system of matter and consciousness that lives within and is sustained by an unbounded field of Life-energy—an infinitely deep ecology that I have called the Meta-universe or the generative ground. (Duane Elgin, *Awakening Earth: Exploring the Evolution of Human Culture and Consciousness,* New York: William Morrow, 1993, p. 274)

In this view, the universe is perceived to be unified in all its parts. Moreover, as physicist David Bohm tells us, it is a holo-movement, a moving or dynamic hologram, in which everything that we experience emerges in each moment from the "implicate order," an energetic but unmanifest domain that harbors the patterns of all creation. Like a hologram, each part of the world we can see and name contains and is contained in each other part, every aspect—butterflies and boulders, bouncing babies and blooming begonias—linked in "undivided wholeness in flowing movement." Fully involved and interconnected, this cosmic organism involves each one of us in absolute intimacy, even identity, with itself. Renowned physicist Erwin Schröedinger echoes this view powerfully when he writes: "Inconceivable as it seems to ordinary reason, you . . . are all in all. Hence this life of yours which you are living is not merely a piece of the entire existence but is in a certain sense the whole" (Erwin Schrödinger, *My View of the World,* p. 21, quoted in Ken Wilber, *Spectrum of Consciousness,* Wheaton, IL: Theosophical, 1977, p. 59).

If we are all in all, a moving hologram of the whole of creation, from what Cosmic Garden of Eden did we emerge? And toward what destiny are we pointed? Perhaps in Jump Time, as the ground of the known shifts beneath our feet, what we need to steady ourselves is nothing short of a new origin myth, an evolutionary tale that takes visionary science as its given.

In the beginning there were and continue to be the Great Gardeners who live in the Metaverse, a vast farm fertile with energy, creativity, intelligence, and love. The Gardeners decide to plant a new garden in a field of the farm's limitless, nested universes. They begin with a infinitesimally tiny seed, a microcosm coded with the energy resources to flower into a richly varied cosmos. So potent is the ground, so ready is the seed, that once planted, it bursts its pod with an explosion of light and energy.

And lo, the infinitesimal seed sprouts into a great tree that holds in its branches a trillion galaxies, each blossoming with a hundred billion or more stars. Whirlwinds of energy swirling through the branches coalesce into biosystems of planetary scale, each home to billions of organisms that balance each other in self-sustaining ecological webs. Nourishing each bud of this immense flowering is the great tree, which links every expression of the garden's unfolding in energetic resonance, such that anything that happens in any part is known instantaneously to the whole.

As the budding life forms of the biospheres complexify, the most advanced among them jump first into awareness of themselves and then into awareness of the Great Gardeners who planted them. Problems that arise at each stage of their growth create opportunities for learning, experimentation, and new expression, leading the advancing ones to deeper and more profound understanding of themselves and their world. As this understanding grows, they develop ways to meet their physical needs with less and less expenditure of energy and resources, so that more and more of their awareness can be devoted to tending the garden of their consciousness and culture. Soon, the winds of the technology they have evolved are crosspollinating the flowers of many places and knowings.

Venturing out to explore the worlds of the very large and the very small, first in their imaginations and then through their technological advances, these adventurous ones come to discover the wonders of the cosmic tree. They begin to understand that all life is engaged in a process of continuous creation and that birth, growth, death, and new birth are all expressions of energy in motion. They come to see that the cosmos both within and without is a living organism, a single unified garden, recreated in its entirety moment by moment by the love and intelligence of the Gardeners, which flows continuously through the Great Tree like nourishing sap.

They discover, further, that along with the knowledge of the Great Tree comes a radical freedom. They know themselves to be free to make mistakes, to face evil, and to experience suffering, for suffering is the inevitable consequence of the great potential of their seeded nature, locked into a still maturing consciousness. Yet, over time, as their scope of vision widens, these beings evolve toward transcending their suffering. As they do, they come to a more and more expansive understanding of who they are and what they yet may be and do.

Knowing at last that all is within all, the totality present in each part and each part fully connected to the whole, these beings—and we are they—move beyond the limited conceptions of the local laws of form and gain access to the very patterns of creation. With this knowledge, they join the Gardeners in their task of planning and planting cosmic gardens and nourishing them with their own intelligence and love. And so the cosmos continues to bloom.

The cosmic story I have told is itself a hologram for living in Jump Time. It reminds us that we, too, are Gardeners who can farm the fields of space/time, the generative ground of our being, creating gardens of consciousness, landscapes filled with the blossoms of our minds and spirits. Tending the gardens of our lives involves a kind of cosmic yoga; we yoke ourselves back to remembering that we are made of the same stuff as the Metaverse from which we continuously arise second by second. We share its body; we are woven into the fabric of its infinite ecology; the productions of our hands and minds are an aspect of its creation and live in eternity. We know ourselves, then, as resonant waves of the original seed, infinite beings who contain in our body-minds the design of creation itself, planted in the field of this particular space-time and sustained by a dynamic flow-through of cosmic energy.

At your core you already know this to be so. Surely, there are times in your life when you understand yourself to be a reality surfer, delightedly riding the waves of creation, mind opened, heart expanded, the Metaverse coursing through you. In such states, you are embraced in co-conscious awareness, no longer knowing or caring where "I" leave off and the rest of reality begins or whether there is any difference. This experience is one of the supreme givens of our nature because the Metaverse in its operational mode is coded into every one of us. The raptures of the deep self are our native equipment, granted us by our cosmic origins. The only requirement is joy and a willingness to say yes to the new epic that dawns, right now, in you and me and those fortunate to be alive in the great today. We are seeds coded with cosmic dreams. Bursting the pods of our containment, we are ready to enter into creative partnership with the Metaverse and to populate our particular corner of space-time with our unique vision and capacity.

A Harvest of Spiritual Practices

What practices might we harvest from the cosmic garden that can help us enter the flow of continuous creation? What might we all do to participate more fully in the vital renaissance that is the emerging spirituality of Jump Time?

Tell the new story of the evolutionary journey, for it is a tale filled with empowering inspiration, a source of hope and change. The story of the origin and growth of the cosmos is being recounted and discussed in many forms, in books, on the Internet, and at international gatherings. One of the most complete and powerful retellings is *The Universe Story* by Thomas Berry and Brian Swimme.

At once the oldest and the newest of stories, the Universe Story heals while it unites. I find that when I tell it in my seminars or guide my students through a reenactment of its stages, it reawakens the memory banks of human history rooted in our cells and psyches and brings an expanded perspective to bear on our work, our arts, and our actions. It replaces alienation with a sense of connectivity; loneliness with an expansive and infinite sense of family; and life-destroying materialism with a spiritual and humane agenda. A myth for our times, it links the personal-particulars of our local existence with the personal-universals of Great Life.

As Mark Steiner, a participant in the Epic of Evolution Internet list, wrote me:

> We need to tell the Epic as a whole as well as to tell the individualized stories of our sun, our planet, our moon, our

> species. We need to tell it in narratives and in dramas. We need to tell it in literal and in metaphorical terms. We need to tell it though art and paintings. We need to tell it in song and in dance. We need to tell it on television and in movie theaters. We need to tell it in comic books, children's books and novels. We need to tell it in street theater, operas, and performance art. We need to tell it in rituals and reenactments. We need to tell it in church services and workshops. We need to tell it in buttons, bumper stickers, and yard signs. We need to tell it with great grandiosity as well as with the most gentle, subtle, even subliminal of tones. We need to so fill the world with its own story that there is no escaping it. We need to so fill our culture with the story that it cannot help but be transformed. Our job may well be to fill our culture with the story so that it seeps deeply enough into our collective consciousness to create a common, contagious unity.

Find a retelling of the Universe Story that appeals to you and make it part of your personal mythology. Tell the story to your children, your friends, your neighbors and colleagues. Keep in mind that there are those whose religious beliefs may cause them to view the story of the universe's evolution as a threat. In telling the new story, stress its universal spiritual dimension without diminishing or disrespecting anyone's creation story. Told as a spiritual practice, the Universe Story has the potential to bring together and celebrate the common heritage of a diverse universe.

Commune with beauty wherever you find it. Spend more time exploring and celebrating the human glories of literature, art, music, dance, and theater and the simple wonders of the natural world. Immersion into beauty wherever you find it calls forth inner beauties and brings to consciousness the budding of new realities and the freshness of a world made new, the esthetics of evolution in action.

Artists catch the currents of the Metaverse and put them into forms that call our emergent selves to heightened awareness. Art makes perception more acute and conception as well. It shakes the mind from its stolid moorings so that you see deeper into the world and time. Active appreciation of nature wakes you up to what is going on around you, heightens your empathy, knits you into a seamless kinship with all living things. Bringing Jump Time knowings to bear, you appreciate the billion-year story that has gone into the making of this rose, that valley, this ocean breeze.

Try it now. Close your eyes for a moment and call up in your imagination three things that enchant you with their beauty—a baby's face, the green hills of Ireland, Michelangelo's David. Visit each in turn, allowing your body and mind to become utterly available to—even merging with—each . . .

Even a moment of such appreciation leads naturally to a state of flow consciousness, the dissolving of boundaries among the knower, the knowledge, and the known. One of my favorite priestesses of this kind of consciousness is the New England poet Emily Dickinson. Caught in the mid-nineteenth century, she managed to bring all time into her own small space, for as she says:

> Behind Me—Dips Eternity—
> Before Me—Immortality—
> Myself—the Term Between—

What happens when you think like Emily and invest each "Term" that you meet—each flower, sunset, prairie, and bee—

with totality? Pursued as a spiritual practice, this kind of thinking leads you to a kind of holonomic knowing, which Emily calls circumference knowing. You allow your mind to wrap itself around its object like a python, but instead of suffocating it, you give it life. You see the before, the after, and the between of things. You catch the glint of glory and the shadows skittering in the corner. Then, like an artist, you burst with words such as no one has ever heard and paint with colors from shores unseen. You dance, like Shiva, the death and resurrection of all, and you comprehend, like a physicist, that everything is implicate and resonant in everything else—"stir a flower and bestir a star." A joyous cosmology becomes apparent, a state in which everything is flowing, pouring, bleeding, seeding, and laughing through to everything and everyone else. Emily Dickinson, that spiritual genius, poured out this revelation in words that melt our very margins:

> Beauty crowds me till I die
> Beauty mercy have on me
> But if I expire today
> Let it be in sight of thee—

In this state, anything on which you focus opens up—projects, problems, relationships, business, governance, metaphysics, even grand designs. You awaken to the wealth of being that is a given of your deepened human condition, and the *aha* experiences keep on coming. You say Yes! to life wherever you find it, abandon whining, welcome and celebrate the springtide of change. Living in the grace of the world's beauty, Grace happens, shift happens, and the mind is prepared to receive Reality in all its many colors.

Decondition old habit patterns that keep us stuck in a state of illusion and forgetfulness. Whether our particular "nonvirtue" is gossip or anger, self-deprecation or toxic thinking, social or business practices that hurt the souls of others, Jump Time spirituality has many methods for bringing such behaviors to consciousness so that they can be replaced by self-nurturing and compassionate ways of being.

Many methods for "changing our minds" begin by helping us to become conscious of the chain of causality that keeps us trapped in a cycle of negative behavior. Eastern mindfulness practices suggest that we notice the arising and passing away of negative impulses, without attaching to them, and by so doing widen the gap between the self and its behaviors so as to provide for an interval of choice. Other schools of spiritual psychology counsel that we practice catching ourselves when we are about to engage in a negative habit and saying, "STOP!" We then change the imagery around the habit pattern by doing or saying or thinking something else. With repeated practice, we learn to self-orchestrate the mind and its callings with greater skill.

Apart from those negative patterns of thought and action accumulated over a lifetime, which are difficult enough to overcome, are the even more entrenched evolutionary debris of leftover archaic attitudes and obsolete programs wired into the oldest part of our brain. However cultured and urbane we may be, when we feel our territory threatened, personally or as a nation, we still regress to fang-toothed screechings and feral crouches on only middling provocation.

The cells and systems of our brains and bodies seem to catch these atavisms like flypaper. Our throwback behaviors are given credence by Bible Belters and behaviorists alike; where one sees fallen nature, the other proclaims the nastiness of our neural fixations. Both exhort us to be dependent on their nostrums for salvation.

Two callings seem to be warring within us. On the one hand, the instinctual drives of habit and conditioning; on the other, the metaphysical calling toward spiritual realization. In this task we have glorious company. The author of the *Prapanna Gita* wrote, around 600 B.C.: "Lord, I know what virtue is, but I cannot practice it; I know what vice is, but I have no power to desist it." Five hundred years later, Saint Paul notoriously complained: "For the good I would, I do not; but the evil which I would not, that I do."

Of what "evils" are we in Jump Time guilty? Matthew Fox, in *Sins of the Spirit, Blessings of the Flesh,* writes eloquently of the sins of our time, the release of which requires us to go more deeply into the shadow realms of our nature. Each is a violation of the sacred, a damning agenda of our capacity for negative creativity through our collective moral and ethical failings. Each tells us much about what needs to be purified spiritually for the world to move to its next stage.

First among the modern deadly sins, Fox writes, is actively "wronging others," adding to the suffering of an already suffering world. But sin, Fox says, can also be passive, such as selfishly choosing not to see, not to hear, not to feel what is happening around us, as we do when we "ignore" our planetary ecological catastrophe. Our habitual ways of thinking can also be sinful, Fox says. In the sin of "reductionism," we oversimplify complex issues between human beings by attributing them all to sexism, racism, or class conflicts. We are also guilty of "dualism," insisting on either/or solutions to complicated and multifaceted dilemmas. Finally, Fox tells us, we moderns sin by a "lack of passion," as when we distance ourselves from the wildness of nature or put out the energetic fire that can propel us and the world forward (New York: Harmony, 1999, pp. 158–59).

The best penance for our personal and collective "sins" may be acts of public service, restorative justice, volunteerism, and philanthropy—acts of kindness on a planetary scale. Take compassionate action and you transcend reductionism and dualism, for you see the larger picture and come to know the richness and complexity of others. Compassionate service engenders passionate concern and sets the life force to flow more strongly in your blood.

Enter the silence or celebrate the fullness by making time for a practice of spiritual connection, logging on to Universe.org, the God Net of consciousness. We know that the universe is a living system of elegant design that seems intent on providing opportunities for learning through every thought, word, and deed. Change perspective through meditation, reflection, or centering, or shift the bandwidth of everyday consciousness to the divine wavelength, and you discover yourself to be the latest flower on the tree of the cosmos, ready to bloom.

One way of sorting through the various methods of contemplative practice available today is to divide them into two broad categories, which we might term "the way in" and "the way out." "The way in," or *via negativa,* is traditionally described as the way of negation. On this path, we retreat progressively from the circumference to the center, clearing the muchness to get to the suchness. "The way out," or *via positiva,* is the way of fullness.

This path takes us out into the world to experience its richness, conscious of developing more and more hooks and eyes to catch the Metaverse.

What's different in Jump Time is that the spiritual technologies at our disposal can be harvested from the whole world: Christian centering prayer, Buddhist mindfulness and visualization practices, African trance dancing, Tantra and sacred sexuality, Native American powwows and sweat lodges, shamanic spirit journeys, Asian martial arts, Jungian dreamwork, as well as, for some, the neomystical study of quantum realities. All of these rework the landscapes of the subliminal mind so that there are channels and riverbeds in which a deeper spiritual consciousness can flow.

In its early stages, meditation on the via negativa tends to require much the same effort as learning to read—that is, a conscious changing of focus. Just as the strange and baffling markings on the page gradually become letters, then words, then sentences, then meaning, so in inward-turning contemplation, we discover a whole new way of apprehending a Reality that is first glimpsed, gradually understood, and finally grasped. As we empty the mind of images and thought, we sink by stages into the great No-thing. Face to face with the substance of all being, the energy of all creation, we discover to our deep joy that the Metaverse was identical to our own consciousness all the time, regardless of how far or how weirdly we sought it outside ourselves.

The study of quantum physics can lead us to much the same conclusion. The evident structures of matter are pared away, revealing Reality in its pure energetic state, which some mystic-minded scientists think is synonymous with the primary energy of consciousness. Brenda Dunne and Robert Jahn of Princeton University, for example, say that consciousness is primary and that quantum events follow. "We do not so much regard quantum mechanics as a metaphor for consciousness, but rather the other way round. We think that the fundamental concepts of quantum mechanics are the fundamental concepts of the human mind" (*Margins of Reality: The Role of Consciousness in the Physical World*, New York: Harcourt Brace, 1987, p. 205).

Quantum physics describes reality as a complementarity between waves and particles. Sometimes everything that exists partakes of one form and sometimes of the other. I suspect that in the contemplations of the via negativa, consciousness moves away from the particle form of the sensate world and plunges into the wave, the fathomless oceanic depths in which Ultimate Consciousness may reside.

The via positiva turns the telescope around and looks through the other end. By active engagement in art, music, dance, movement, theater, and high play, our perception is extended to the world without and the world within, our senses sharpened so that we bathe in multisensory delight. Through heightening our awareness, bringing more and more content into consciousness, and opening to the entire spectrum of emotions, sensations, and ideas, we come to the realization that all things are interdependent and part of the living life of the Metaverse. We celebrate the particle in order to catch the wave.

Choose whatever forms of innering or expanding practice appeal to you and make time and space for them in your life. Whether you "go out" or "go in" doesn't matter; the important thing is to get going!

Find a community to support your spiritual practice. "Practice-oriented spirituality," observes Princeton sociology professor Robert Wuthnow, is "a way of imposing discipline on personal explorations" ("Returning to Practice," *Noetic Sciences Review,*

August–November 1999, p. 37). I suspect that in the future even traditional churches and synagogues will adopt an eclectic range of practices. Human potentials seminars, business spirituality groups, men's and women's circles, sanghas and ashrams, retreat centers and body-and-soul conferences, even Internet mail lists, cater to every style and flavor of practice. Their proliferation is further evidence that the spiritual zeit is getting geisty.

If you can't find an existing group that suits your taste, you can always start your own. I spoke earlier (chapter 3) about the importance of teaching-learning communities and gave suggestions for starting them. Such communities are particularly important to spiritual practice. Gather a group of like-minded seekers and make a pact to support each other's regular spiritual endeavors. As Buddhists explain, the three jewels of spiritual practice are Buddha, the teacher; dharma, the teachings; and sangha, the community of fellow practitioners. Without the support of sangha, they tell us, the spiritual path can be lonely and difficult, indeed!

A New Springtime of Spirit

As more and more people adopt practices like these, signs of new growth are everywhere. In contrast to the wasteland of government, grassroots movements sprout, connecting fields of ideas and greening the social agenda with greater community responsibility and inventiveness. Millions of cultural creatives are adopting voluntary simplicity and putting economics back where it belongs, as a satellite to the soul of culture, thus restoring the social balance. A new appreciation and celebration of our relationship to Nature is rising, rewriting our covenant with the

Earth and acknowledging that we humans are her steward and partner, not her master. Aging baby boomers are acknowledging that our elderly are critical to the health of the planet, true citizens of Jump Time who can deal wisely and creatively with planetary complexity because they have lived long enough to develop the necessary depth and simplicity.

People are responding to the stress of current issues by going beyond themselves. Many are learning skills they never thought to have, inspired by an undeniable inner urge to take on heroically creative tasks they never thought to do.

Joan in Massachusetts, a respected neurobiologist, is leading a revolution in science to bring together brain research and spirituality.

Francis, a male nurse and former monk, redeems intractable patients in a California schizophrenic ward with the sweetness of his nature and the depth of his compassion.

Teresa, a New York oncologist and a mystic, uses her spiritual presence and healing gifts to alleviate the dread and fear of cancer patients.

Vendana, a physicist and ecologist in Delhi, writes and acts to combat corporate piracy of the world's botanical heritage.

Catherine, a Connecticut documentary filmmaker, creates programs for public television that illumine the human spirit.

Helmut, a stockbroker in Berlin, organizes relief efforts to bring new hope to the children of war.

In their own ways, these and many more like them are citizens of the Metaverse, models of Spirit in action.

The world mind, it seems, is forging a new container for its spiritual seekers. Whether it is a new religion or the blending of the best of the old ones, whether it is more universal forms of collective worship or a general intensification of private spiritual practice, something unprecedented is brewing in the Earth's spiritual continuum. Perhaps the most hopeful sign of Jump Time is just this: the grand company of mystic-minded adventurers, bent on exploring every room in the many-mansioned House of the Holy.

CONCLUSION

SIGNPOSTS ON THE ROAD TO THE FUTURE

T. S. Eliot tells us in "The Four Quartets" that a person who gets on a train and reads a newspaper is not the same person who gets off. In the same way, you who began this book last week or last month and persisted to the end are not the same person you were on page 1. A book like this is like a train; it takes you through a variety of terrains, some racing by so fast that you are continually saying, "What was that?" And doubtless other experiences took place along the track beyond the book that added to your appreciation of the passing landscape.

So here we are, changelings all, at the end of the line. But before you get off, the conductor of this particular journey would like to offer some suggestions for how to proceed when you leave the station, how you might best put to use what we've experienced together.

At the beginning of the journey, I spoke of the five forces that serve as the motive power for Jump Time. Taken together, they are signposts that provide directions for finding your way through the woods of a world in which everything is changing, everyone is in transition. They may even point the way to the New Story for the world as well as to things you can do to help that story emerge. Let's look at each force in turn and read its message.

The Repatterning of Human Nature

I sometimes say that as human beings we are not flaky but snowflaky—so different are we from one another. Thus the repatterning of human nature takes many different forms. One way of discovering your form is to ask yourself:

- In what directions do I need to grow to develop a fuller use of my potential? What might I do to develop them?
- What qualities, skills, talents are dormant in me that could be expressed more fully?
- How would developing these qualities better the world and give me the courage and vision to take on the tasks and challenges of Jump Time?

In the course of the book, I have mentioned a number of human potentials that can help configure the possible human for the Jump Time world: access to the entelechy or essential self; intuition; the ability to self-orchestrate across states of consciousness; synesthesia—the capacity to hear colors or smell words; entry into the virtual worlds of the creative imagination; turning on the polyphrenic self; journeying as a cybernaut and as a psychenaut; becoming a cultural fusionary; traveling one's own mystic path; and many others. Undoubtedly, you already have some of these abilities; others are in you as latencies. These qualities are aspects of the nascent human self that has been in preparation for millennia but seems to need the stress and ferment of Jump Time to be quickened into form. Developing these abilities also helps counter the millennial blues many of us are experiencing because of so much change and the loss of traditional meaning and security. When one is "cooking on more burners," the despairing self tends to take a lesser role in the family of selves.

High tech requires the balance of high touch if we are not to become the pets of our machines. By "high touch," you may recall, I mean not just human qualities such as those just listed above, but anything that enhances one's embodiment in flesh and bone, sense and sensibility. Through high-touch activities, we come to inhabit ourselves again, in ways that we may have lost during the long industrial night of the body. Then, too, listening to the wisdom of our body makes us available to the wisdom of the Earth. When we enhance our senses and perceptions, we become more accessible to the medley of other cultural forms—ethnic and fusionary alike. We are more likely to play in the fields of the world's repast—in food or music or literature or ways of being—if we are tuned in through enriched perception rather than shut out because of ancient xenophobias.

Until you get used to this new way of being in the world, high touch takes active pursuit. Seek out and interact with people and groups outside your normal sphere. Make an effort to listen so that you can hear what people intend, feel, and experience rather than simply what they say. Look for the essence underneath the apparent picture and use that insight to communicate with others in a way they can understand.

What links these high-touch capacities for self-repatterning is that each implies a different relationship to time. Jump Time asks that we intend to be in time and that we make each moment conscious. When we live with awareness and active attention, old barriers to perception and communication collapse. Developing a new relationship to time also helps us with the dismay many of us feel when we survey our harried lives. We all have so much to do that to add time to our already overcrowded days to work consciously on personal repatterning seems impossible—unless we can find a way of working within an alternate time modality. The most useful alternate modality is what I call subjective time. When we are in subjective time, a few minutes of clock time seems much longer; in fact, it is all the time you need.

Training yourself in subjective time is very simple. Give yourself a minute of clock time. Close your eyes and see how many experiences, travels, projects you can imagine within one minute. As you continue to practice this exercise, gradually increase the interval until you are experiencing five minutes of clock time as equal to an hour or more of subjective time. Then use your five-minute excursions into subjective time to rehearse and improve any skill you wish to develop. Most people find that their abilities improve much more rapidly when they practice in subjective time. Perhaps this is because in subjective time we tend to think in images, the mind capable of synthesizing and selecting, sometimes doing the work of months in minutes. Using subjective time, you will have plenty of time to do the added work of Jump Time.

To lose too much time in Jump Time is to lose one's opportunity to play a role in the greatest transition drama the world has seen. Thus we need to become time players and not time victims. Use your developing inner senses to dream, create, and carry out your vision of yourself as an active creator and participant in the challenge of making a better world—at home, in your workplace, in your community. I have found in my research that people who have the gift of follow-through and thus consistently get things done invariably use the creative skills and imageries of the inner senses that emerge within the rubric of subjective time. They complete their tasks in the outer world because they are continually inspired and energized by a passion for the possible that is coming from the inner world. As they cultivate the gift of directed imagination, their projects grow from inside out rather than the other way around.

It is also important in Jump Time to learn new skills—two a year at least—so that you are developing from the outside in as well. The brain grows through use. As you acquire a particular skill, you catalyze an assemblage of related abilities. Thus learning to use a computer for writing and as a worldwide communication tool brings with it mental dexterity, the capacity to hold in mind a weave of ideas, an intercontinental savoir faire, and appreciation for the world of information and for passionate communication. Learning a second or third language fosters the extended cultural identity so necessary in Jump Time. The dexterity developed through mastering a new art form or skill—pottery or bread-making or cross-country skiing—grants access to another order of intelligence. Rehearse your new skill in subjective time, calling for advice and instruction on that part of your polyphrenic self that is expert in the new capacity: the Artist, Master Chef, or Athlete.

Most important is staying power with the gifts received and acquired. Drop in before you start to drop out. Access your finer nature, your entelechy, and be directed in passion and purpose. If at all possible, join or start a teaching-learning community that keeps you growing along with like-minded friends and allies committed to discovering and nurturing each other's hidden talents. Then together, move out in the community, see what needs vision and renewal and, using your new capacities, go out and do it. It is quite possible that before too long you will be recruited for tasks that go beyond the community. Such has been the experience of many who have agreed to the terms of participating in Jump Time. For just as technology has rescaled us to planetary proportions, the Earth within gives us the corollary wisdom to know our place and tasks in this era of stupendous change. The mutations of present time seem to be asking no less than the immediate transmutation of ourselves.

The Regenesis of Society

Predicting how society will recast itself is like dropping a thousand numbered Ping-Pong balls down the stairway and guessing which one will be first to reach bottom. I am reminded of the predictions of H. G. Wells, thought in his time to be a great prophet of the future. In 1939 he wrote of his conviction that

> after the Second World War [there will remain only] a patchwork of staggering governments ruling over a welter of steadily

> increasing social disorganization. It will be the Dark Ages over again. . . . There will be a return to primitive homemade weapons, nonmechanical transport, a new age, not of innocence, yet of illiteracy, and slow, feeble and less lethal mischiefs will return to the world.

Wells was mistaken because he could not see beyond his own pessimism at a time when Hitler was looming over Europe, fascism was spreading, the world was just emerging from the Depression, and Japan's militant reign of conquest was reaching its apogee. The author of *The Time Machine* clearly had never taken a trip in his own literary invention. But what if we were able to send that machine back to Wells in 1939 and use it to bring him forward? What would he make of the current whirlwind of change, the daily advent of technological wonders, life in the NOW zone, and a world run more by corporations than by governments?

For that matter, what would we see if we could travel forward sixty or more years, say, to the 2060s? Would we discover a world ruled under some form of planetary governance, with nations as states within a world federation? Would the stress of overpopulation have driven humanity out into space as seedlings of the stars? Would human lives be vastly longer, our cells linked to machines, our minds capable of merging with vast knowledge banks? Would science have shifted its paradigm from machines to living systems? Would we have a win-win world where almost everyone had access to physical and spiritual nourishment, education, and opportunity? Or would we have the blistered, busted world that William Gibson portrays in *Neuromancer*—with cyber-jocks jacked in to virtual realities as they skitter across appliance-littered landscapes?

Whatever will be there in the future owes everything to what we do today.

- How, then, can we Jump Time players contribute toward the regenesis of society?
- What can we do to help create a social order in service to life rather than to economic forces?
- What is the world that we would like our great grandchildren to inherit?

These questions are impelling a growing number of social pioneers working on the frontiers of a new society. What they are doing is perhaps the most important social movement the world has ever known. Economist David Korten hails them in his powerful book *The Post-Corporate World.*

> These determined pioneers are creating new political parties and movements, strengthening their communities, deepening their spiritual practice, discovering the joyous liberation of voluntary simplicity, building networks of locally rooted businesses, certifying socially and environmentally responsible products, restoring forests and watersheds, promoting public transportation and defining urban growth boundaries, serving as peacemakers between hostile groups, advancing organic agriculture, practicing holistic health, directing their investments to socially responsible businesses, organizing recycling campaigns, and demanding that trade agreements protect the rights of people and the environment. (San Francisco: Berrett-Koehler, 1999, p. 3)

These pioneers are everywhere, in every social and ethnic group, race, and profession. Perhaps you are among them. Fed up with elitist leadership and distant bureaucracies, you have discovered what Korten calls the powerful potential of direct democracy and are ready to take responsibility for the health and well-being of yourself, your families, your communities, and the planet.

Further, you are charged by the knowledge that the Internet is returning to individuals much of the authority lost to institutions during the explosion of industrialization. Suddenly you have more control over information and resources, a development that is already playing havoc with traditional top-down government, business, even religion. You are part of a revolution in autonomy as well as responsibility. In the world of the noosphere, densely interconnected communication networks of people who cherish their communities and care deeply about life on this planet are creating something never seen before: a meta-sphere of governance. Grounded and responsive to each individual, this movement involves every area of the Earth in a conscious, self-organizing, life-serving planetary process. The Earth herself is becoming a vast teaching-learning community, a new order of democratic biology, as individuals and groups learn as they participate and create together a new arena for social evolution.

All of this requires a move toward *politeia,* a Greek word much juicier than *politics,* which derives from it. Politeia is civic society carried to its utmost. It implies that all free citizens are actively involved in creating and maintaining their community, empowering each other to further their common social agenda. In Jump Time especially, it requires the plumbing of our emotional depths, the manifesting of our minds' best thinking, and above all the courage to follow through. As psychologist Arnold Mindell says, "Creating freedom, community, and viable relationships has its price. It costs time and courage to learn how to sit in the fire of diversity. It means staying centered in the heat of trouble."

As faith, hope, and love are to religion, so politeia is to politics. Politeia recalls the ways in which Deganawidah and his allies created the Longhouse of Longhouses in the five Iroquois nations. In our time, a politeia is a kind of culture virtually identical with what Paul Ray and others have called the integral culture. Duane Elgin and Colleen LeDrew track this culture in their "Global Paradigm Report":

> An integral culture . . . seeks to integrate all the parts of our lives: inner and outer, masculine and feminine, personal and global, intuitive and rational, and many more. The hallmark of this integral culture in an intention to integrate—to consciously bridge differences, connect people, celebrate diversity, harmonize efforts, and discover higher common ground. With its inclusive and reconciling nature, an integral culture takes a whole-systems approach and offers hope in a world facing deep ecological, social, and spiritual crisis. ("Global Paradigm Report: Tracking the Shift Under Way," *YES! A Journal of Positive Futures,* Winter 1997, p. 19)

Here are seven forms within which a politeia can operate toward the regenesis of society. Many of these ideas are already being practiced in forward-thinking communities around the world, but they are, unfortunately, still the exception rather than the norm. If the possible society we all dream of is to be made manifest, they must become second nature.

1. A ***politeia of participation*** provides all members of society with opportunity to influence political and economic insti-

tutions affecting their lives and fosters personal responsibility to fulfill these tasks. No one in Jump Time is innocent of responsibility; each of us has a role to play. Think about what your role might be. Perhaps your community needs a safe place for teenagers to hang out, or the local independent bookstore needs support. Perhaps schools, or senior centers, or battered women's centers need volunteers. Imagine it, write about it, read about it, talk to encouraging friends about it. Get some allies and do something about it, one step at a time. Or, as a Seattle environmental activist who had just celebrated her hundredth birthday said after telling the story of her remarkable life, "You do what you can. And then you do some more."

2. A ***politeia of rediscovery*** rekindles spontaneous generosity and neighborliness and honors the capacities of others. In this world of human snowflakes, a neighborhood is an extraordinary collection of ingenious, eccentric, brilliant, skillful, weird, and wonderful types. One of the reasons people watch sitcoms is that the actors and script writers play out the lives of typical families and communities in only slightly exaggerated form. What if you turned off the TV and turned on your neighborhood instead? You might find stars of human possibility waiting to be discovered. The greatest of all human potentials is the one that recognizes and evokes other people to bloom. Seek out ways of encouraging each person's capacity by improving and deepening the processes of human life: our birthing, our parenting, our nutrition, our health, our fitness, our family and community life, our education, our arts, our sciences, our ways of growing old, and our ways of dying beautifully. These concerns make us fully human.

3. A ***politeia of creativity*** activates the artistic process to

recharge our imaginations. You and your allies might organize a festival of way-out ideas, new inventions, creative visions, modes of expression. Get together a group to paint a community mural, landscape the town square, start a craft co-op, bring local businesses and schools together to design and build a playground. There's nothing the human imagination working in community cannot accomplish.

4. A ***politeia of healing*** moves us beyond the polarities of left versus right, or us against them, and promotes cooperation, understanding, and networks of mutual aid. I spoke in chapter 6 about the work of Leah Green, who helped facilitate compassionate listening between Israelis and Palestinians. What if you were to create a meeting place where, once a month, conflicting groups—skateboarding teens and local police, public utility companies and conservationists—could talk to each other? Before the session, act out a compassionate listening dialogue to model the technique, making your interaction heartfelt and even funny so that people lose their resistance and want to try. You'll find that when people start to dialogue in a spirit of fairness and deep listening, critical issues can be addressed with wisdom and integrity.

5. A ***politeia of celebration*** encourages music, songs, humor, dances, rituals, and myths of possibility to be played out, performed, and celebrated. Building community in the new millennium requires that we create social theater to tell the New Story of a world in transition. A woman I know in a small town in rural Georgia took the stories of her area's local history and created a pageant called *Swamp Gravy* in which everyone in the community participated. What if we each became a local impresario for Jump Time talent, from preschoolers to seniors? If we

each encouraged schools, community and civic groups, churches, and neighborhoods to get involved, we might trigger a renaissance of theatrical and musical engagement in our communities as a fertile seedbed for new ideas and vision.

6. A ***politeia of hope*** encourages an attitude of wonder and astonishment before what can be. It treats the problems and scarcities before us as opportunities to clarify what is really important. In this, we might well take a lesson from the Iroquois people who, before their council meetings, spoke their praise and gratitude for everything, naming their blessings with words and songs of wonder. Think how different it might be if your local school board meeting began with gratitude for the community's blessings and with hopeful declarations rather than with partisanship, discouragement, or resentment. Hope give us energy and motivation and avails the human mind of its finest treasures.

The Breakdown of the Membrane

I cough, and a rice farmer in Cambodia sneezes. Well, maybe not exactly, but the metaphor is useful. The Earth is a living system, interdependent in all its parts, just as our physical body is interconnected in all its cells and organs. Today this interdependence has become endemic. Membranes of every variety—ecological, psychological, and sociological—are breaking down or opening up, allowing a streaming of content that compounds the complexity of all and everything. The electronic networks give us a world without walls; television invades all private spaces, filling people half a world apart with each other's sights and sounds and feelings; the migrant tides of humans swim, sail, fly, and walk across old borders and refuse to go back. The DNA exchanges that result from the ensuing matings pulse with hybrid vigor.

There is no going back. Countries like Japan that have been successful in keeping out migration "to keep the race pure" are experiencing the diminishment of their population and labor force, so much so that Japan's prime minister announced, only half jokingly, that at current rates of decline, by 2060 Japan would have zero population! To be successful, cultures must collapse their walls and let their contents stream forth to form new multicultural units. Open-membrane technology creates the world car, the world computer, even the world shoe, with parts drawn from all over Asia, South America, and Europe. "Made in the USA" now tends to mean that a bunch of international parts have been assembled in Tennessee.

But this story of breakdown has a deeper meaning that some may miss. It has to do with the nature of membranes, particularly the most basic, the membrane of the cell. The cell membrane is a gate that lets things in and out, a dynamic physical barrier flexible enough to accommodate the processes of movement and growth. Proteins in the membrane function like little antennae, receiving stations for signals from the environment. They pick up information and make changes in the membrane so that the cell can regularize or transform itself.

In the evolutionary scheme, cell survival was linked to its ability to increase the surface area of its membrane. As the environment became more complex, cell membranes grew to accommodate new information. But this growth had an upper limit. When the membrane of a single cell was no longer large enough to protect the cell from the environment, the cell experienced an evolutionary jump. Over a period of 2.5 billion years and as a result of huge planetary stresses, single-celled organisms joined with others to increase their collective chances of survival. As individual cells came together, they created ways to share the information in their DNA, so that each cell in a ten-cell community had access to the DNA in the other nine. Eventually, this shared DNA became wrapped in a nucleus more or less in the center. The centralization of genetic information led to cells becoming specialized and to their joining together in even more complex forms of cooperation. Multicellular organisms merged into cellular communities, whose membranes interpenetrated and exchanged information. From these evolved plants and animals, which are essentially colonies of interconnected communities.

Human society followed a similar pattern in its growth from families to tribes to larger social organisms, until now we stand on the verge of becoming co-conscious organisms of planetary proportions. This evolution is both our gift and our curse. Old fundamentalisms rise like the thickening walls of cells that refuse the jump into multicellular form. Individuals and institutions that react to the stress of Jump Time with xenophobia and withdrawal end in stasis—gigantic amoebas, failed, angry, paranoid.

Let's take this analogy one step further. Say civilization is a colony of cells. Governments are the nucleus of each cell, centers where the DNA strands of history are held. DNA tends to be conservative and stable; its aim is to insure the replication of the past, the preservation of the status quo. Could this be why politicians so often apply yesterday's solutions to today's problems? Moreover, the nucleus does not provide for communication between internal and external environments. Think of the insensitivity and lack of responsiveness characteristic of bureaucracies, and you'll see what I mean. Quite apart from governments and their nucleated strongholds, human beings have created a cell membrane called culture, which takes in information and converts it into useful forms. The streaming, permeable membrane of culture flows around the homeostasis of government, exchanging information in vital ways, despite fortresses of defense, Star Wars weapons, and the entrenchments of preservationist economic policy.

As a species we are essentially at a place as critical to our future as when amoebas faced the jump to multicellular life forms. Our survival seems to demand that nucleated structures break down, that individuals and institutions come together into new colonies that share a larger common membrane. Whether these new multicellular entities are neighborhoods in which households share responsibility for each other's well-being or Earth bioregions that ignore national borders because of common ecological imperatives, there's no question that the movement is toward new collective forms, which carry with them the need for new levels of trust and cooperation.

This metaphor of the cell membrane can help us track our personal level of permeability. Imagine that you are a cell. The ways you interact with the world outside yourself—your entire feeling, thinking, interfacing, cultural self—is a membrane that encases the nucleus of your self-preserving, conservative, habit-driven self. Think for a moment about what it would be like to locate your consciousness in the membrane rather than in the nucleus. From this perspective, with your consciousness spread out along the permeable borders of your being, ask yourself the following questions.

- With whom and with what am I willing to "multicell"—to jump into larger associations, networks, friendships?
- Who and what makes me withdraw, even panic?

• Do these people or situations threaten only the "nucleated" or conservative aspects of my being?

• Or is there something in me that is drawn, shaking and quaking, to risking the encounter and moving into a more inclusive reality?

Sometimes it helps to prepare yourself for the jump to multicellular living with an inner exercise, one that stretches the membrane of your imagination to a more natural feeling for inclusiveness, like the following.

Close your eyes and think of two people with whom you feel a natural "multicellular" affinity. Let yourself feel a connection with them. Then imagine that each of these people is able to send out similar feeling lines of connection to two others, and these in turn to two others. In a few moments, virtually everyone in the world will be connected. Now extend your web of connections to animals, plants, creatures of all kinds, until every sentient being is connected. Then ask the Earth herself to do the same with two other planets, and they with two more, until all the planets in the galaxy are connected. And from our galaxy, connect to two more, and two more, until the entire universe is connected. All are enclosed within the membrane of the mind of God or Being, and from Being, radiant intelligence and love is sent back through the weave of all things and beings to you. And from you to the heart of all Being, and from the heart of all Being back to you. Stay in this resonant and connected state for a while.

Now consider again those people or associations from whom you shrink and those toward whom you are drawn to jump into a larger membrane of commitment. That urge to jump, to join, to connect, to break down the membranes that divide us, is the lure of becoming, the drive to evolve toward being an enriched, engaged, complex participant in a larger social organism, a scout on the frontiers of Jump Time.

The Breakthrough of the Depths

An Aboriginal woman once took me on a walk to some of the sacred places near and around Uluru, the great stone mountain in the center of Australia, sacred to her people. As we passed each site, she told me hilarious stories about what the ancestral spirits had done in the great timeless time known as the Dreamtime to create that place and how the souls of Aboriginal people were waiting there to enter whatever pregnant woman passed by, regardless of race. "Whitefella thinks they own this land," she laughed. "But now many new whitefella babies are really blackfella. Their ma caught one of our spirits. Now whitefella catch our dreaming."

The following week a psychiatrist in Sydney told me about a phenomenon he was seeing in his practice. It seemed that a number of his patients, most of English and Irish descent, were dreaming of Aboriginal symbols, especially *chirungas*, shields that portray symbolically the journey of an ancestral spirit. "Bingo!" thought I, recalling the words of the Aboriginal woman. The psychiatrist went on to say that until his patients tracked the meaning of these dream images, they were filled with nameless dread, as if a critical part of themselves had gone missing.

However one chooses to interpret this story, it seems to be part of a phenomenon that crosscultural travelers are encountering more and more frequently. With space shrinking and time warping, the depths of the psyche—ours and the Earth's—are breaking through, and the collective unconscious is having a field day.

I suspect this is a further consequence of the thinning of the local membranes that have kept us nucleated and separated from each other, as well as of the growth of the collective membrane, through which we are sharing the content of our psyches in unexpected ways. We catch not just each other's moods and thoughts but dreams and cultures and even memories as well.

Sometimes I think that the old hippies are having their revenge, and LSD has finally been put into water supplies around the world. Dreams are becoming wilder, visions abound, people everywhere are getting "downloads" of the "truth," and prophets, channelers, and spiritual master rascals are two for a quarter. The mania of the current spiritual scene may be the necessary carnival aspect of the breakthrough of the depths. In the ancient Greek theater, roguish satyr plays preceded and followed the great dramas of the soul of Aeschylus, Sophocles, and Euripides; they took the edge off the powerful psychological catharsis that attends soul at its turning.

All over the planet, what had been hidden in the collective psyche is now emerging, more festive and feisty than ever. The questing journeys of millions of seekers in the current spiritual renaissance are sourced in this collective rising. Saint Francis once said that "what we are looking for is Who is looking." Perhaps it is our innermost psyche, the "nature that lies within," the inner mirror of Great Nature, which has pushed the universe at large along on its evolutionary journey, that is the Who calling us to be more than we ever thought we could be.

- How do we meet and respond to this breakthrough in ways that do not blow us out but help us engage as cocreators of this rising wave of Spirit?

• How can we distinguish between mystic and mad perceptions?

• How can we use those aspects of the self that are kin to the rising depths to access and help others avail themselves of the extraordinary spiritual power that is now at hand?

As we have seen, consciousness has the innate capacity to tune and modulate to different domains. It moves easily through four levels of awareness: the physical, the psychological, the archetypal, and the mythic. It is equally at home in the realm of the local and personal This Is Me, the mythic and archetypal We Are, and the unitive and universal I Am. By changing consciousness, we gain the ability to experience the profound patterns of the universe.

When our sensory and psychological base is more comprehensive—seeing the world and each other in all the ways we can, polyphrenic, tuned in to the entelechy—we gain the inner tools for social artistry. Our psychological makeup has more source hooks in it, is less traumatized by past experience, is more capacious, and we feel extended into a multidimensional universe.

Then, when we bring these enhanced skills and perceptions to the mythic level of consciousness, we breathe another atmosphere and partake of a larger universe with regard to perception, time, space, dimensionality, and possibility. When we live mythically or consciously enact an archetype, we seem to pass into a reality in which what had been latent becomes overt. What we think of as "paranormal" phenomena—clairvoyance, extraordinary strength, out-of-body travel—may belong to another system within the universe that is mediated by archetypes. Rapid healing, for example, can be the result of the healer dissolving his or her local self and being filled with an archetypal or sacred image—not I but Christ through me doeth the work. By bridging oneself between here and the Greater There, one enters into archetypal dimensions that may contain blueprints of a greater possibility, the primal stuff for social and creative change. Archetypal space-time may also contain the optimal template of a person's health and well-being. The job of the healer is to call that template back into consciousness so that it can work upon a malfunctioning body or mind, tuning here, correcting there. Extended to the social sphere, the work of the healer is to use the breakthrough of the depths to catch a vision of the Great Patterns of Possibility. Then, filled with the passion for the possible, one does the work of carrying out the vision in the larger community.

Enlarged and enhanced in this way, we discover spiritual creativity working through us as the instruments through which the larger order of Reality enters into time. Practically, there are many things we can do. For example, organize a church and synagogue round robin in which congregations take turns visiting each other. Reach out beyond the Judeo-Christian community to interact with Buddhists, Hindus, Muslims, and Native Americans in your area. If you belong to a largely white church, get to know a black church, and vice versa. Share projects. Cooperate on solving community problems. Celebrate the values that link you despite your differences.

In India, when a person needs special support or encouragement, he might say to a neighbor, "Will you be god for me for today? Hold me in prayer. Give me the benefit of your advice and counsel." Playing "god" for each other in a humorous and benign way calls forth our deepest wisdom and our finest listening skills.

When we offer each other the support of Spirit, we quicken the godseeds in all of us.

You might try this exercise imaginatively first, or enact it with a trusted friend who agrees to an experiment in the breakthrough of the depths. Sit quietly and follow your breathing for a few minutes. Allow your local conditions and habits to release their hold and fall away. Only the vibrant intelligence that you have winnowed from years of living remains. If your friend is not present physically, then imagine her sitting before you and hear what she might say about her condition and needs. Then look at your friend from the state of I Am. No matter how many problems you may have at the local This Is Me level, make the shift to "god" consciousness now. In sacred time and space, you are moving into a realm in which you transcend all despair and access all knowing.

You are now the godded being, looking at your friend from a place of limitless love and empathy. You may even feel yourself sending her high re-creative energy. Speak, if it seems right, about releasing old patterns and of the higher pattern that is hers to receive—a pattern that heals while it makes whole. Allow that pattern to emerge in the mystery that is there between you. If you have particular insights, offer them. Stay in resonance for a while until you feel your I Am consciousness connecting with your friend's I Am, godself to godself. Speak to each other or in your imagination about what you have felt and understood. End by giving thanks for the Grace that is present when Spirit breaks through and moves through us all.

The Evolutionary Pulse from Earth and Universe

"Hello out there? Is anybody home?" Is the Earth just a cosmic rock barnacled with organisms called humans? Are the stars just cold fires harboring other cosmic rocks? I prefer to think not, and probably you don't either. Instead I see a universe alive in all its parts, communicating with all its parts, receptive and responsive, a living field of life and consciousness. In the introduction, I suggested that the motive force of Jump Time is the pulse from Earth and Universe calling us to become partners in creation, stewards of the Earth's well-being, conscious participants in the epic of evolution. The Earth's story, as we have seen, is bigger than all of us yet very intimate, for it desires our engagement, our love, and our commitment.

People everywhere began feeling the current call to transformation when they first saw the Earth from outer space. Somehow that picture made all humans into family again, with one another and with the Earth. In a sense, Jump Time began with that picture. For it was in looking at our planet that we knew ourselves looked at. In feeling, we knew ourselves felt. The Earth herself began to pulse through, penetrating the asphalt avenues of urban towns and the abstract halls of parliaments and government agencies. It was then, in the late 1960s, that the ecology movement took off, as well as other movements for greater social responsibility. People hugged trees, kissed the soil, and took steps to save the whales and protect endangered creatures. On the human side, they become peacemakers, seeking new pathways to righteousness, health, and creative power for the disenfranchised peoples of the planet. It was as if the Earth herself was having a say in the creation of our collective New Story.

Call up in your mind's eye now that picture of the Earth as seen from outer space, and ask yourself:

- What is my place in this New Story?
- How might I feel that pulse rising from the Earth and learn to move to its rhythm?

Why not do something to answer these questions right now? Put down the book, go outside, and lie on the Earth. Get in touch with its rhythms, is dynamic equilibrium, its perfect poise. Let your borders dissolve. Feel the nutrients of Earth and its wisdom pass through the membrane of your organism, its higher pattern pulsing through, not just for another phase of growth, but for the deepening of the whole planet. Become matrixed and mothered. Feel yourself back in the womb of creation, Earth's fetus, being rewoven into new forms and new connectivity. Feel yourself opening to a larger and more comprehensive continent of spirit.

We have come to that stage where the real work of humanity begins. This is the time and place where we partner Creation in the recreation of ourselves, in the restoration of the biosphere, and in the assuming of a new kind of culture—what we might term a culture of kindness, in which we live daily life reconnected and recharged by the Source so as to become liberated in our inventiveness and engaged in the world and our tasks. We come into what may be termed flow consciousness. Mystics have always written about this state. But recently, scientists have used similar language to describe the way the universe works.

The universe as we presently know it is expanding rapidly. It consists of at least fifty billion galaxies, each of which has in the neighborhood of one hundred to four hundred billion stars and the Universe only knows how many planets with intelligent life on them. Space is not empty. In fact, even a complete vacuum contains tremendous amounts of background energy. The majority of the cosmos is dark matter; that is, for each particle in the tangible universe, there may be something like a trillion shadow particles in universes in other dimensions. You and I may be replicated, with variations of course, in an infinite number of parallel universes.

Further, theorists tell us that the cosmos is an integrated system that is deeply and totally unified in some mysterious way. Things that seem separate from each other are really connected through influences that transcend the limitations of ordinary time and space. Immense levels of energy flow through the universe constantly and, what is more, continuously regenerate it. You are new at every second. Or as physicist Brian Swimme puts it, "The universe emerges out of an all-nourishing abyss not only 15 billion years ago but in every moment." Thus, everything in the universe is a flowing movement that arises with everything else, moment by moment, in a process of continuous regeneration. And what's more, the cosmos itself may be a hologram, such that each part of the flow as it continually recreates itself contains in miniature the whole blooming picture.

If the cosmos is a hologram, any change anywhere results in a change everywhere. All is within all, the totality present within each part, and each part fully connected to the whole. And if each part of the universe contains the immensity of everything, perhaps our universe is itself enfolded in a Meta-Universe, beyond our local laws of form, that connects us instantaneously to a vast cosmic system in which higher creative laws predominate. When our cosmos blossomed into existence from an infinitely small pinpoint some fifteen billion years ago, it emerged out of an infinitely deep domain of Being's vast intelligence, creativity, and energy. Perhaps this domain is the Meta-Universe, the ultimate, generative ground of our being. It provides the nutrients and light of the Life Force that continuously sustains our cosmos as a unified organism. So self-reflective and co-conscious is our universe in all its parts that at a certain level of evolution, its organisms can achieve awareness of the Meta-Universe, even tap into its fecundity, as well as adding their productions to the Mind of the Whole.

Allow yourself to marvel at the immensity of this picture for a moment, and then ask yourself:

- Where does my life fit into the consciousness of a universal reality?
- What can I do to play my part?

As we become conscious of our participation in an immensity of consciousness, we begin to realize that each human body-mind is constructed so as to have the capacity to receive from the pulses of the Universe the knowledge and the power to make the changes that can save the world.

In this universe of ours, from the tiniest particles to the Galaxies of galaxies, everything is rhythm, pulse, beat, music. The music of the spheres is no mere metaphor, and its cadences are the stuff that supports life and all its becomings. Get in touch with the rhythm, which is your essence, your essential self, and the rest unfolds in the music that is meant to be. We are each of us melodies in the Divine symphony, and we can join the orchestration of the whole with the music of our minds.

Are the stars out tonight? Then have I got something for you!

Find yourself a spectacular star-worthy piece of music, like the last section of Beethoven's Seventh Symphony. Go outside. Put on the music. Raise your eyes and your arms to the heavens. Conduct the stars, knowing each point of light to be a player in your orchestra. Signal various stars or star groups to take the lead in parts of the symphony, just as an orchestra conductor might give the downbeat to the woodwinds or the violins. Put your whole body into the music. Allow the sky to respond until you no longer know whether you are the conductor or the conducted. When the music ends, continue conducting, this time the music of the spheres. Know your body and mind to be a grand chorale in which everything is frequency, vibration, resonance, love. Thrill to the polyphony that is Jump Time.

Sentient Publications, LLC publishes books on cultural creativity, experimental education, transformative spirituality, holistic health, new science, and ecology, approached from an integral viewpoint. Our authors are intensely interested in exploring the nature of life from fresh perspectives, addressing life's great questions, and fostering the full expression of the human potential. Sentient Publications' books arise from the spirit of inquiry and the richness of the inherent dialogue between writer and reader.

We are very interested in hearing from our readers. To direct suggestions or comments to us, or to be added to our mailing list, please contact:

SENTIENT PUBLICATIONS, LLC

1113 Spruce Street
Boulder, CO 80302
303.443.2188
contact@sentientpublications.com
www.sentientpublications.com

16428766R00162

Made in the USA
Lexington, KY
26 July 2012